PRACTICAL GAS METAL AND FLUX CORED ARC WELDING

PRACTICAL GAS METAL AND FLUX CORED ARC WELDING

Mike Gellerman

Prentice Hall
Upper Saddle River, New Jersey *Columbus, Ohio*

Library of Congress Cataloging-in-Publication Data
Gellerman, Mike.
Practical gas metal and flux cored arc welding / Mike Gellerman.
p. cm.
Includes index.
ISBN 0-13-639238-5
1. Gas metal arc welding. 2. Electric welding. I. Title.
TK4660.G384 1999
671.5'22—dc21 98-23219
CIP

Cover photo: © Super Stock
Editor: Ed Francis
Production Editor: Stephen C. Robb
Design Coordinator: Karrie M. Converse
Cover Designer: Rod Harris
Production Manager: Patricia A. Tonneman
Production Supervision: Carlisle Publishers Services
Marketing Manager: Danny Hoyt

This book was set in Century and Olive by Carlisle Communications, Ltd. and was printed and bound by Banta Company. The cover was printed by Banta Company.

Simon & Schuster/A Viacom Company
Upper Saddle River, New Jersey 07458

Printed in the United States of America

10 9 8 7 6 5 4 3 2 1

ISBN: 0-13-639238-5

Prentice-Hall International (UK) Limited, *London*
Prentice-Hall of Australia Pty. Limited, *Sydney*
Prentice-Hall Canada, Inc., *Toronto*
Prentice-Hall Hispanoamericana, S. A., *Mexico*
Prentice-Hall of India Private Limited, *New Delhi*
Prentice-Hall of Japan, Inc., *Tokyo*
Simon & Schuster Asia Pte. Ltd., *Singapore*
Editora Prentice-Hall do Brasil, Ltda., *Rio de Janeiro*

Preface

The main objective of this book is to help readers develop the abilities to do welding efficiently and safely. The learning material is recommended for use in industrial arts classes, adult education, and vocational and apprenticeship programs, and for all students who want to learn basic welding. This book is written to meet some of the specifications set forth in the American Welding Society's *D1.1 Structural Welding Code—Steel.*

Questions are included throughout. Besides providing a thorough content review, many of the questions have been developed to encourage readers to apply their skills in mathematics and writing as well as problem solving. Carefully following the information provided in this book will enhance readers' enjoyment of welding and improve their ability to make quality welds.

This book has an easy-to-understand format for use with an instructor. By following the step-by-step approach in each unit, readers will increase their comprehension of the material and master the basics. And the best news is that almost any person who is healthy and has good vision and hand-to-eye coordination can learn basic welding. Students motivated by a conscientious effort to answer the review questions and to complete the welding exercises as often as necessary to meet the criteria established for the class will achieve success in welding.

Welding has its own set of terms, and unfamiliar words will be introduced throughout this book. Words that appear in **bold type** can be found in the glossary at the end of this book. Readers should look them up and try not to skip over words that are unfamiliar. Finally, readers should make every effort to follow the equipment manufacturers' recommendations for safety and complete their own mental "safety checklist" each time they set up to do welding. Remember that success at welding will depend on the time and the effort spent to accomplish the given tasks.

Although specific attention has been given to particular terms and their definitions, the limitations of a text of this size devoted primarily to welding skills make it necessary to simplify many of the explanations. A comprehensive discussion of information introduced in this text can be found in technical publications produced by the American Welding Society.

ACKNOWLEDGMENTS

The author gratefully acknowledges the following reviewers for their insightful suggestions: Joe Beckhan, Hobart Institute of Welding, ITW Welding Product Group; Charles Brown, San Jacinto College; Andrew Burke, Monroe County Community College; Dave Hoffman, Fox Valley Technical College; Carl Matricardi, Gwinnett Technical Institute; Ron Whitman, Texas State Technical College.

The author also thanks Dan Maynard, welding instructor at Wisconsin Indianhead Technical College, for his assistance on this project, and Cliff Anderson at Northeast Wisconsin Technical College, who did the welding for Figure 5-5 on common welding problems. The author also thanks Carlisle Publishers Services, copy editor Carey Lange, and graphic artist Andrew Van Osdol. Finally, the author thanks the staff at Prentice Hall, especially Executive Editor Ed Francis, who understood the author's vision for this book.

Contents

UNIT 1 Welding Theory 1

The Process of Welding 1
Definition of Welding 1
Definition of Gas Metal and Flux Cored Arc Welding 1
Description of Gas Metal and Flux Cored Arc Welding 4
Electric Power 7
Basic Electric Theory 7
Input Current 7
Output Current 9

UNIT 2 Welding Equipment 13

Safety in Welding 13
Safety First 13
Power Sources 15
Choice of Welding Machines 15
Duty Cycle 18
Five Types of Power Sources 19
Equipment Necessary for Welding 23
Welding Equipment 23
Power Source 24
Feeder System 26
Gun 27
Workpiece Connection 30
Filler Metal 31
Shielding Gas 32
Safety Equipment 34
Safety Glasses 34
Ear Plugs 34
Respirators 35
Welding Helmet 36
Welding Gloves 38
Clothing 38

Cap *39*
Tools *39*
Ventilation *41*

UNIT 3 Welding Materials 49

Metals for Welding 49
Properties of Metal *49*
Steelmaking *51*
Low-Carbon Steel *53*
Identifying Low-Carbon Steel *54*
Steel Plate 55
Preparation for Welding *55*
Welding Electrodes 59
Composition and Identification *59*

UNIT 4 Fundamentals of Welding 65

Act of Welding 65
Beginning the Arc *65*
Travel Angle *65*
Work Angle *66*
Direction of Travel *66*
Electrode Extension *67*
Travel Speed *68*
Metal Transfers 69
Short Circuiting Transfer *69*
Globular Transfer *69*
Spray-Arc Transfer *70*
Two Methods of Flux Cored Arc Welding *70*
Suggestions to Improve Welding Technique *71*

UNIT 5 Common Welding Problems 75

The Weld Bead 75
Two Basic Methods *75*
Failure of the Weldment *77*
Slag *79*
Cracking *81*
Arc Blow *83*

UNIT 6 Welding Exercises 89

Two Positions for Welding 89
Flat Position and Horizontal Position *89*
Lab Exercise One 91
Short Circuiting Transfer (GMAW-S): Laying a Bead *91*
Equipment and Material *91*
Welding Parameters *91*
Directions *92*
Lab Exercise Two 92
Short Circuiting Transfer (GMAW-S): Padding Plate in Flat Position *92*
Surfacing Weld *92*
Equipment and Material *93*
Welding Parameters *93*
Directions *93*

Lab Exercise Three 94
Short Circuiting Transfer (GMAW-S): Corner Joint in Flat Position 94
Fillet Weld 94
Equipment and Material 94
Welding Parameters 94
Directions 95
Lab Exercise Four 95
Short Circuiting Transfer (GMAW-S): Lap Joint Horizontal Position 95
Fillet Weld 95
Equipment and Material 96
Welding Parameters 96
Directions 96
Lab Exercise Five 96
Short Circuiting Transfer (GMAW-S): Lap Joint on Sheet Steel in Horizontal Position 96
Fillet Weld 96
Equipment and Material 97
Welding Parameters 97
Directions 97
Lab Exercise Six 98
Short Circuiting Transfer (GMAW-S): Butt Joint on Sheet Steel in Flat Position 98
Square-Groove Weld 98
Equipment and Material 98
Welding Parameters 98
Directions 99
Lab Exercise Seven 99
Short Circuiting Transfer (GMAW-S): Butt Joint on Sheet Steel in Horizontal Position 99
Square-Groove Weld 99
Equipment and Material 99
Welding Parameters 99
Directions 100
Lab Exercise Eight 100
Short Circuiting Transfer (GMAW-S): Lap Joint on Sheet Steel in Vertical Position 100
Fillet Weld 100
Equipment and Material 101
Welding Parameters 101
Directions 101
Lab Exercise Nine 102
Short Circuiting Transfer (GMAW-S): Butt Joint on Sheet Steel in Vertical Position 102
Square-Groove Weld 102
Equipment and Material 102
Welding Parameters 102
Directions 102
Lab Exercise Ten 103
Spray-Arc Transfer: Corner Joint in Flat Position 103
Fillet Weld 103
Equipment and Material 104
Welding Parameters 104
Directions 104

Lab Exercise Eleven 105
Spray-Arc Transfer: Tee Joint in Horizontal Position 105
Fillet Weld 105
Equipment and Material 105
Welding Parameters 105
Directions 106
Lab Exercise Twelve 106
Spray-Arc Transfer: Butt Joint 106
Equipment and Material 106
Welding Parameters 106
Directions 107
Lab Exercise Thirteen 108
Flux Cored Arc Welding (FCAW-G): Corner Joint in Flat Position 108
Fillet Weld 108
Equipment and Material 108
Welding Parameters 108
Directions 108
Lab Exercise Fourteen 109
Flux Cored Arc Welding (FCAW-G): Butt Joint in Flat Position 109
Square-Groove Weld with Backing 109
Equipment and Material 109
Welding Parameters 110
Directions 110

UNIT 7 Joint Design 113

Five Basic Joints 113
Choice of Joint Design 113
Fillet Weld 113
Groove Welds 114
Weld Size 117
Other Welds 117
Selecting a Joint Design 119

UNIT 8 Electrode Wires and Shielding Gases 125

Choice of Electrodes 125
Types of Wire Electrodes 126
Electrode Sizes 129
Storage of Electrode Wires 130
Variety of Electrode Wires 130
Choice of Shielding Gases 131

UNIT 9 Heat and Distortion 139

Effects of Welding 139
Expansion and Contraction 139
Distortion and the Amount of Welding 141
Setup for Welding 143

UNIT 10 The Welding Project 149

Repair and Fabrication 149
Repair 149

Fabrication 150
Tools 151
Layout 154

APPENDIX 157

Guided Bend Test 157
Introduction 157
Qualifications of Testing 158
Preparing Coupons for Testing 159
Testing the Coupons 161
Lab Exercise One 161
Short Circuiting Transfer (GMAW-S): Tee Joint in Vertical Position 161
Fillet Weld 161
Equipment and Material 161
Welding Parameters 161
Directions 162
Lab Exercise Two 163
Spray-Arc Transfer (GMAW-S): Butt Joint Meets AWS Structural Welding Code 163
V-Groove Weld with Backing in Vertical Position 163
Equipment and Material 163
Welding Parameters 163
Directions 163
Lab Exercise Three 164
Short Circuiting Transfer (GMAW-S): Tee Joint in Overhead Position 164
Fillet Weld 164
Equipment and Material 164
Welding Parameters 165
Directions 165

GLOSSARY 167

INDEX 179

UNIT 1

Welding Theory

THE PROCESS OF WELDING

Definition of Welding

Several processes can join materials by **welding,** as shown in Figure 1-1. A basic definition of welding applied to the subjects discussed in this book is a process of joining material together by heat. Welding can be likened to carpentry. Both woodworking and metalworking provide the opportunity to put ideas into practice and require a plan that involves measuring, cutting to size, and fastening together. Learning how to weld provides the opportunity to fabricate or repair products by using a small, reasonably priced welding **power source** (Figure 1-2).

Welding is one method of joining metals different from **riveting, bolting, soldering,** or **brazing** (Figure 1-3). In **gas metal** or **flux cored arc welding,** the **filler metal** mixes with the **base metal.** Welding is the only method of joining metals in which the filler metal, or holding material, mixes completely with the base metal. The filler metal combines with the base metal to form the **weld.** The weld takes on the characteristics of both the filler metal and the base metal. In most instances, the **joint** becomes stronger than the original base metal.

Definition of Gas Metal and Flux Cored Arc Welding

Gas metal arc welding (GMAW) is a welding process that uses an electric arc to generate heat (Figure 1-4). The **welding arc** is formed between the base metal and a filler metal electrode which is continuously fed from a solid or **tubular** spool of wire. The **electrode,** or filler metal, flows into the pool of liquid metal which is called the **weld pool.** The weld pool is a mixture of filler metal and the base metal, or that material being welded. A gas

Arc Welding Processes

atomic hydrogen welding (AHW)
bare metal arc welding (BMAW)
carbon arc welding (CAW)
carbon arc welding-gas (CAW-G)
carbon arc welding-shielded (CAW-S)
carbon arc welding-twin (CAW-T)
electrogas welding (EGW)
flux cored arc welding (FCAW)
gas metal arc welding (GMAW)
gas metal arc welding-pulsed arc (GMAW-P)
gas metal arc welding-short circuiting arc (GMAW-S)
gas tungsten arc welding (GTAW)
gas tungsten arc welding-pulsed arc (GTAW-P)
plasma arc welding (PAW)
shielded metal arc welding (SMAW)
stud arc welding (SW)
submerged arc welding (SAW)
submerged arc welding-series (SAW-S)

FIGURE 1-1 Chart of welding processes. Gas metal and flux cored arc welding are two of eighteen arc welding processes.

FIGURE 1-2 Miller Electric is one U.S. manufacturer of welding power sources.

shield protects the electrode and the weld pool from contamination by the surrounding air. The **shielding gas,** among other attributes, stabilizes the arc and keeps **oxygen** and **nitrogen** in the air away from the weld pool. The completed weld joins metal in a rigid structure and is the result of melting base metal and filler metal together in a joint.

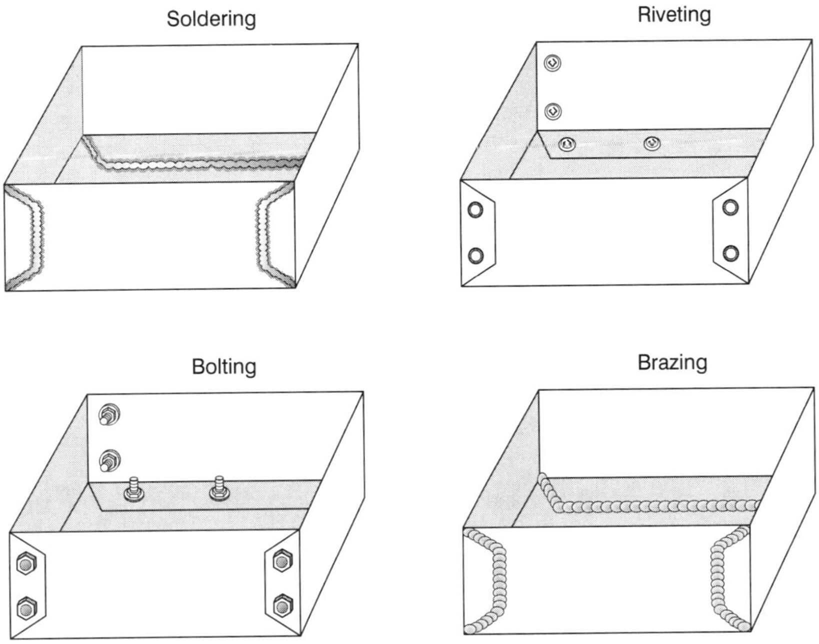

FIGURE 1-3 Soldering, riveting, bolting, and brazing are four other methods of joining metals.

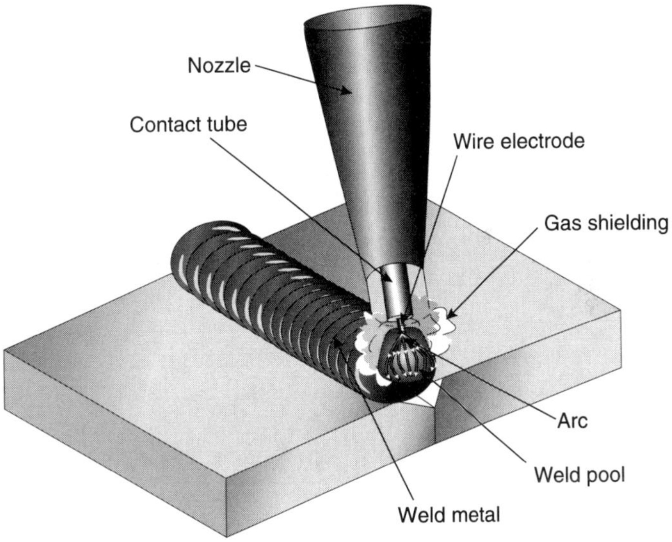

FIGURE 1-4 Wire electrode in GMAW melts into the weld pool protected by a gas shielding.

Flux cored arc welding (FCAW) is another welding process that uses an electric arc to generate heat. The welding arc is formed between the base metal and a filler metal electrode which is continuously fed from a spool of tubular wire. FCAW as a **self-shielding** method uses a shielding gas formed by a powdered flux contained within the tubular wire (Figure 1-5). FCAW as a gas-shielding method uses an additional external shielding gas supplied for protection of the weld pool from contamination by the surrounding air (Figure 1-6). The welding application will

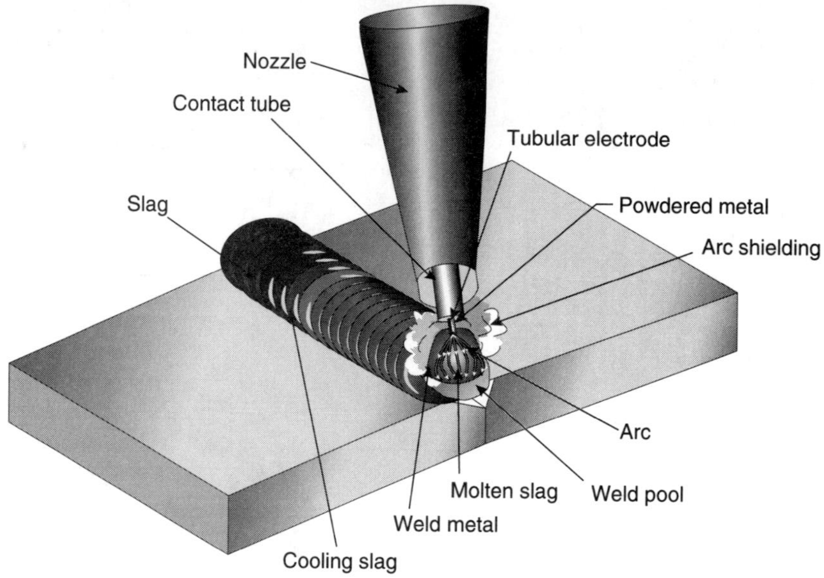

FIGURE 1-5 Tubular electrode in FCAW melts into the weld pool protected by slag as the weld cools.

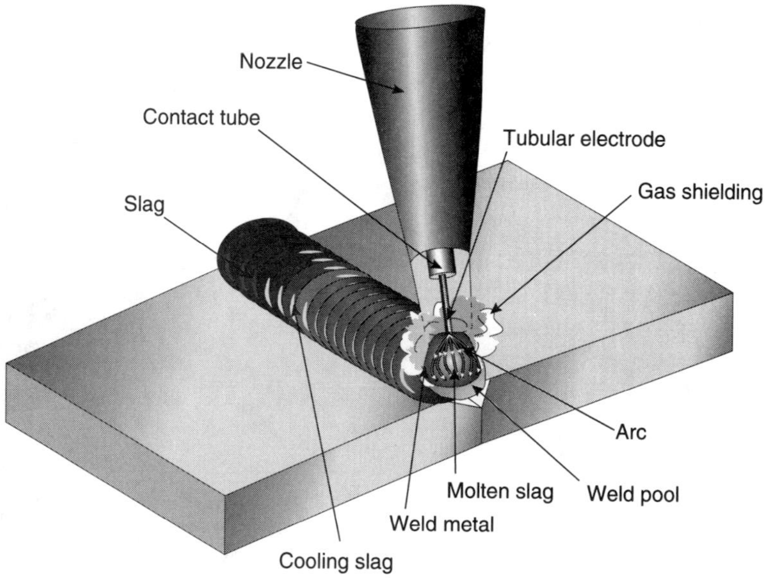

FIGURE 1-6 Tubular electrode in FCAW melts into the weld pool with the added protection of gas shielding.

determine the choice of filler metal. The design of the filler metal will determine whether an external shielding gas is required.

Description of Gas Metal and Flux Cored Arc Welding

Gas metal arc welding (GMAW) is either a **semiautomatic** or an **automatic** process that can save time and money, because the welding can be done without first stopping to remove **slag** from

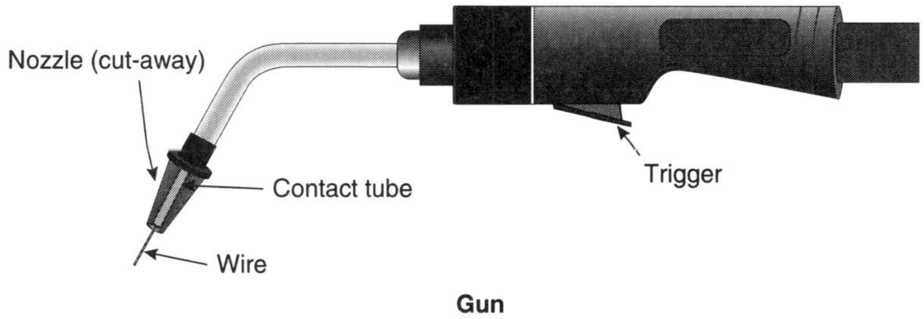

FIGURE 1-7 Major gun parts that feed a continuous wire electrode.

each **weld bead** or to frequently replace an electrode stub with another electrode. Equipment for this welding process, however, is more complicated and more expensive to operate than for shielded metal arc welding. This equipment should be protected from rain, snow, and dusty conditions. Outdoor welding subjects the shielding gas to the effects of the wind, thus requiring the construction of a wind screen or other means of maintaining adequate shielding.

Flux cored arc welding (FCAW) is either a semiautomatic or an automatic process similar to **shielded metal arc welding (SMAW),** in that the slag that forms over the weld bead must be removed before making another pass. In this process the filler metal is continuously fed through the arc into the weld pool. Its uses include the welding of **carbon, low-alloy,** and **stainless steels.** Although FCAW uses small diameter wires on thin gauge **sheet metal,** this process has more extensive application for welding **plate,** in which large amounts of deposited filler metal are required.

Both GMAW and FCAW require the **welder** to pull the trigger during the welding process. The filler metal electrode is pushed out the **gun** through the **contact tube** (Figure 1-7). Upon touching the base metal, resistance of the electric current generates heat in the electrode. Depending on the process, metal filler is transferred from the electrode either when the electrode is in contact with the base metal or across the arc in the electric circuit. As the welding continues, the distance of the contact tube from the base metal (through raising and lowering the gun) changes the resistance in the wire; however, the arc **voltage** remains constant with every momentary change in the **arc length.** For example, when the contact tube is brought slightly closer to the base metal, the rate of the electrode burn-off rises by an increase in welding amperes to bring the arc back to its desired length, thus maintaining a **constant potential.**

In most applications GMAW and FCAW use a constant potential output. This avoids the large drop in voltage that occurs in initiating the arc during the shielded metal arc welding process (SMAW). SMAW uses **constant current** electric **output,** which means the welder is unable to maintain a precise arc length, and must therefore constantly adjust for a changing voltage while welding. With the gas metal or flux cored arc welding process, control of welding **parameters** is set on the power source first. The wire electrode comes off the spool at a constant rate. The wire electrode

FIGURE 1-8 Voltage level and wire feed speed controls.

is controlled by adjusting the speed of the wire feed and the arc length is controlled by adjusting the voltage level (Figure 1-8).

Learning how to make **quality** welds by either GMAW or FCAW should take less time than learning the skills of shielded metal arc welding. With SMAW the student must spend more time learning how to manipulate the electrode and control the arc length while feeding the filler metal into the weld pool at a constant rate. Students learning GMAW and FCAW discover both the rate the wire electrode is transferred into the weld pool and the arc length cannot be altered during welding. Students now have the important task of learning how to adjust parameters on the power source for their particular welding jobs.

ARC FLASH

Gas metal and flux cored arc welding equipment should be covered when not in use if maintained in a dusty environment.

SAFETY REMINDER

The shop should be equipped with one or more fire extinguishers placed in designated locations. Know the classification of each fire extinguisher for correct application. See Unit 2 for a discussion on the four types of fire extinguishers.

REVIEW

1. How is welding related to carpentry?
2. What happens to the electrode during the welding process?
3. What is the purpose of a weld?
4. Describe what happens in the arc.
5. What is the purpose of a shielding gas?

ELECTRIC POWER

Basic Electric Theory

Either gas metal arc welding (GMAW) or flux cored arc welding (FCAW) begins by turning on the electric power. ***Current, voltage,*** and ***amperage*** are three terms at the foundation of electrical theory (Figure 1-9). Current is flow. Amperage is measurement of current flow. Voltage is measurement of the force or pressure causing the flow of electricity.

Using a garden hose to illustrate, current is like the amount of water and voltage is like the pressure of the water in the hose. Amperage is the amount of water flowing through the hose measured in gallons. If the amperage setting is too low, like a dripping hose with low water pressure, there is not enough current provided. The result can be a weld with incomplete **root penetration** into the joint. If the amperage setting is too high, like a fire hose with pressure too high to water a garden, excessive current can cause **melt-thru** of the base metal through the backside.

The welder must learn to regulate the voltage and the amperage to make a quality weld. The voltage level is adjusted on the power source to control the arc length. The amperage level is adjusted by setting the speed of the wire feed. The voltage and the amperage become the output current in the arc during the process of welding.

Input Current

Alternating current (AC) and **direct current (DC)** are used in electric welding. *Alternating* (AC) is the term described simply as the back and forth motion of current through zero. Current changes directions 120 times per second within an electric circuit. Each back and forth motion of current through 360° is called a cycle (Figure 1-10). The standard **frequency** in the United States is 60 cycles per second. AC supplies the **input current** to power familiar household appliances such as the electric shaver and the hairdryer.

Direct current (DC) is the second variation of electric power. Unlike AC, DC is current flow in one direction. DC is used to operate electrical systems on automobiles, motor homes, and power boats. Most important, DC is the output current used to generate the heat for gas metal arc welding and flux cored arc welding. A

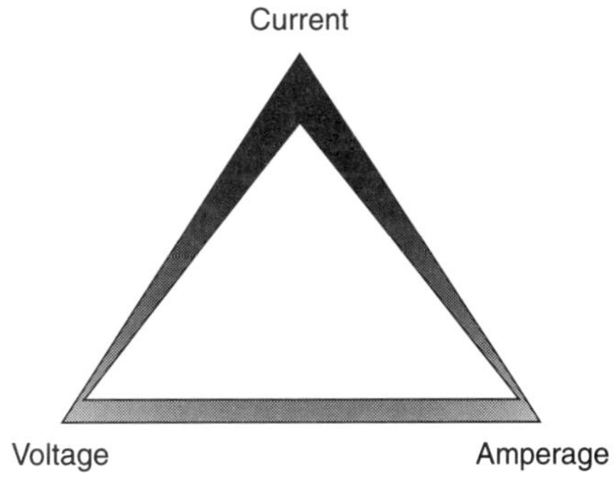

FIGURE 1-9 Three basic terms of electricity.

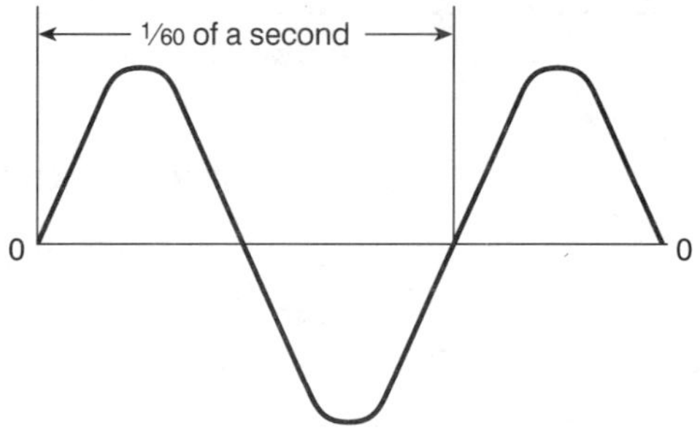

FIGURE 1-10 The wave cycle shows zero where the arc exits and restarts with every change of directions, which is 120 times per second. AC choice of welding current is not as stable as DC.

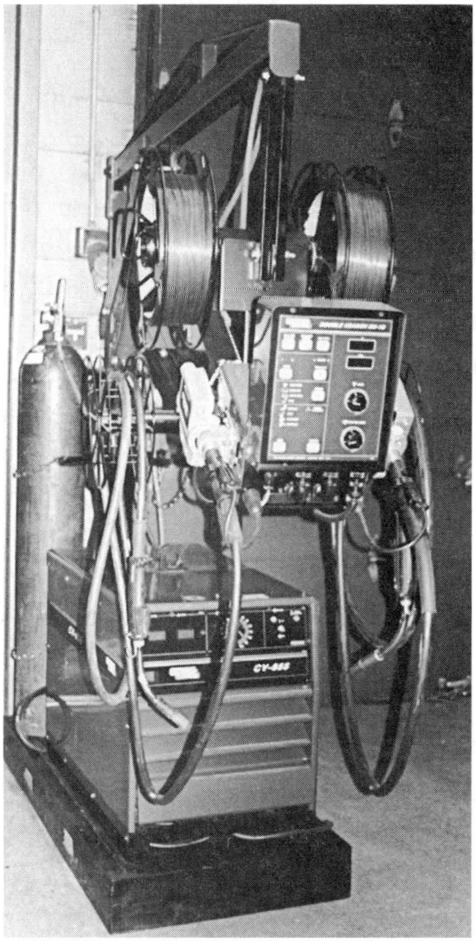

FIGURE 1-11 Lincoln Electric manufactures power sources for gas metal and flux cored arc welding.

rectifier is a device on a power source used to convert AC input electric current to DC output electric current (Figure 1-11).

The power source for welding utilizes a transformer to convert high voltage and low amperage AC input current into low voltage and high amperage output current necessary for welding. The AC is converted to DC output by use of a rectifier. Remember the power source used for welding should only be serviced by a qualified electrician.

SAFETY REMINDER

Be sure the shop is clear of **volatile** and **flammable** materials before making any attempt at welding.

REVIEW

1. Use the following words in a paragraph: *current, voltage, amperage, flow, pressure,* and *measurement.*
2. How does input current differ from output current?
3. What is AC or alternating current?
4. What is DC or direct current?
5. What is the purpose of a rectifier?

Output Current

Welding has three available choices of current: alternating current, **direct current electrode negative,** and **direct current electrode positive.** Alternating current (AC) is not readily adaptable for gas metal and flux cored arc welding in most applications. Using AC involves overcoming the instability of the arc when changing directions from the negative to the positive. Direct current electrode negative (DCEN) is required for the application of some electrodes. With DCEN, the welding gun is attached to the negative terminal and the **workpiece connection** is attached to the positive terminal on the power source. DCEN is used for some FCAW self-shielding applications. Direct current electrode positive (DCEP) is the first choice of current commonly used in external gas-shielded welding applications. With DCEP the gun is attached by its welding cable to the positive terminal and the workpiece connection is attached to the negative terminal on the power source. The current choice of DCEP can be changed to DCEN by switching the welding cable to the negative terminal and the workpiece connection to the positive terminal.

REVIEW

Explain the relationship between the following sets of terms:

a. DCEP and gun
b. DCEN and workpiece connection
c. AC and arc stability
d. DCEN and root penetration

Lab Exercise

Examine a power source in the shop.

a. Explain how to change the current from DCEP to DCEN.
b. Write down the information of any warning labels on the power source provided for your safety.

REVIEW QUESTIONS FOR UNIT 1

Multiple Choice

Choose the *best* answer to complete the statement.

1. Welding:
   a. Is useful in repair.
   b. Is used in fabrication of products.
   c. Joins metal.
   d. Can mean the application of heat.
   e. Is all of the above.
2. Base metal is:
   a. The filler metal.
   b. Nonmagnetic.
   c. The weld.
   d. The material being welded.
   e. None of the above.
3. The weld is:
   a. The filler metal.
   b. The base metal.
   c. The same as soldering or brazing.
   d. Answers a. and b. above.
4. Filler metal is:
   a. The electrode.
   b. Nonmagnetic.
   c. Powdered.
   d. Contaminated.
   e. None of the above.
5. Shielding gas:
   a. Can be an external gas.
   b. Can form from the flux.
   c. Protects the weld pool.
   d. Protects the electrode.
   e. Can be all of the above.
6. The weld pool:
   a. Is liquid metal.
   b. Needs protection of shielding gas.
   c. Is a combination of base metal and filler metal.
   d. Is the welding arc.
   e. Is a., b., and c. above.
   f. Is a., b., and d. above.
7. Gas metal arc welding involves:
   a. A wire filler metal.
   b. A gas shielding provided by an external flux.
   c. A shielding gas that stabilizes the arc.
   d. All of the above.
   e. Answers a. and b. above.
   f. Answers a., b., and c. above.
8. Flux cored arc welding:
   a. Is a slag producing process.
   b. Uses a solid wire filler metal.
   c. Can require an external shielding gas.
   d. Uses a tubular wire filler metal.
   e. Is a., b., and d. above.
   f. Is a., c., and d. above.
   g. Is b. and d. above.
9. DCEP or DCEN:
   a. Is alternating current.
   b. Is current flow in one direction.

c. Requires a rectifier to convert AC.
d. Powers familiar household appliances.
e. Is all of the above.
f. Is b. and c. above.
g. Is a. and c. above.

Short Answer

1. Explain why gas metal arc welding is easier to learn than shielded metal arc welding.
2. Explain the difference between constant potential and constant current.
3. List and define the three terms at the foundation of electrical theory.
4. Describe what the voltage level controls on GMAW and FCAW power sources.
5. Explain how the amperage level is adjusted on GMAW and FCAW power sources.
6. Explain the difference between AC and DC.
7. Define DCEN.
8. Describe what the power source does to high voltage and low amperage current.
9. Describe what can happen if the amperage setting is too low.
10. Describe what happens when the contact tube is brought closer to the base metal.

ADDITIONAL ACTIVITIES

Riddles

1. I am the special equipment supplying the current and voltage for welding.
2. I am the added metal that forms in the joint.
3. I cause intense heat between the base metal and the electrode.
4. I protect the electrode and the weld pool from contamination.
5. I can be solid or tubular.
6. I am the distance of the electrode from the weld pool.
7. I am the welding variables set on the power source before welding begins.
8. I flow in one direction only.

Writing Skills

1. Write a paragraph using the following terms: *weld pool, shielding gas, oxygen, filler metal,* and *base metal.*
2. Write a paragraph in which you explain why you are taking a course in welding.

Math

Common Fractions and Decimals

Understanding the relationship between decimals and fractions is especially important for welders. Wire sizes are given in both fractions and decimals.

0.1 = one tenth	0.01 = one hundredth
0.001 = one thousandth	0.0001 = one ten-thousandth
1/10 = one tenth	1/100 = one hundredth
1/1000 = one thousandth	1/10000 = one ten-thousandth

Solve for fractions.

Examples: 0.3 = ? = 3/10 and 0.059 = ? = 59/1000

1. 0.5 = ? **2.** 0.7 = ? **3.** 0.9 = ? **4.** 0.05 = ?

5. 0.01 = ? **6.** 0.37 = ? **7.** 0.68 =? **8.** 0.003 = ?

9. 0.047 = ? **10.** 0.0007 = ?

Solve for fractions and reduce answers.

Example: 0.4375 = ? = 4375/10000 = 875/2000 = 175/400 = 7/16

11. 0.3125 = ? **12.** 0.1875 = ? **13.** 0.5625 = ?

Solve for decimals.

Examples: 3/10 = ? = 0.3 and 77/1000 = ? = 0.077

14. 7/10 = ? **15.** 4/10 = ? **16.** 57/100 = ?

17. 35/1000 = ? **18.** 45/1000 = ? **19.** 75/100 = ?

20. 63/1000 = ? **21.** 3/16 = ? **22.** 1/8 = ?

23. 7/16 = ?

Solve for the following expressions in words.

Example: 0.057 = ? = fifty-seven thousandths

24. 0.030 = ? **25.** 0.0002 = ? **26.** 0.0105 = ?

$$\text{Watts} = \text{amperes} \times \text{volts}$$
or
$$P = I \times E$$

27. 1/16-inch flux cored wire electrode at 190 amperes with an arc voltage of 21 is how many watts?

28. 3/32-inch flux cored wire electrode at 300 amperes with an arc voltage of 28 is how many watts?

29. 0.030-inch solid wire electrode at 260 amperes with an arc voltage of 25 is how many watts?

Puzzlers

1. Jim is looking for a power source to mount on the back of his truck for on-site maintenance welding of logging equipment. What must he consider before choosing a constant potential power source for use of GMAW or FCAW?
2. What are the factors in selecting a power source? How do you choose between a feed unit that can be separated and carried to a remote location and a feed unit that cannot be removed from the power source?

UNIT 2

Welding Equipment

SAFETY IN WELDING

Safety First

Safe electric welding practices will help prevent injuries. Develop an attitude about safety in everything you do. Serious injury can happen to even experienced welders when they become careless. Think safety.

Gas metal arc welding (GMAW) and flux cored arc welding (FCAW) processes produce sparks hot enough to start a fire. A welder must know the type of fire extinguisher required to put out a given fire. Following are the four main types of fire extinguishers.

- Type A fire extinguishers are for use on combustible materials such as cardboard, wood, and paper (Figure 2-1). Type A is recognized by the symbol of a green triangle with the letter ***A*** in the center.
- Type B fire extinguishers are for use on volatile substances including gas, oils, paints, and solvents (Figure 2-2). Type B is recognized by the symbol of a red square with the letter ***B*** in the center.
- Type C fire extinguishers are for use on electrical fires such as in fuse boxes, motors, and wiring (Figure 2-3). Type C is recognized by the symbol of a blue circle with the letter ***C*** in the center.
- Type D fire extinguishers are for use on metal fires that include magnesium, titanium, and zinc (Figure 2-4). Type D is recognized by the symbol of a yellow star with the letter ***D*** in the center.

Smoke, light, heat, sparks, and shock are five areas of concern with gas metal arc welding (GMAW) and flux cored arc welding (FCAW). The smoke given off by welding can contain fumes and gases that may be harmful if inhaled over time. **Ultraviolet light**

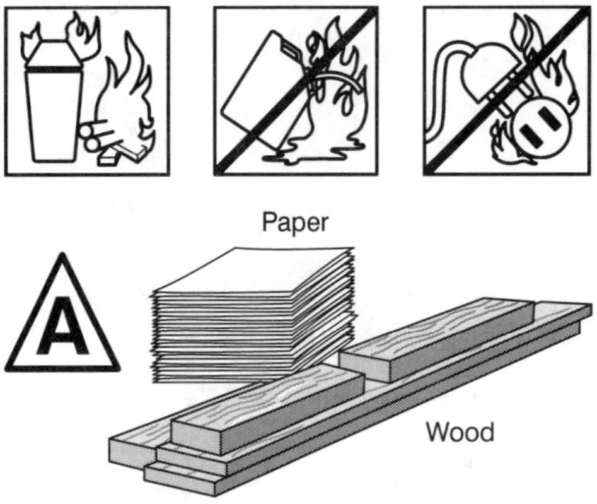

FIGURE 2-1 Type A fire extinguisher. The symbol of a green triangle is for combustible materials such as wood and paper.

FIGURE 2-2 Type B fire extinguisher. The symbol of a red square is for volatile substances such as gas, oil, and paint.

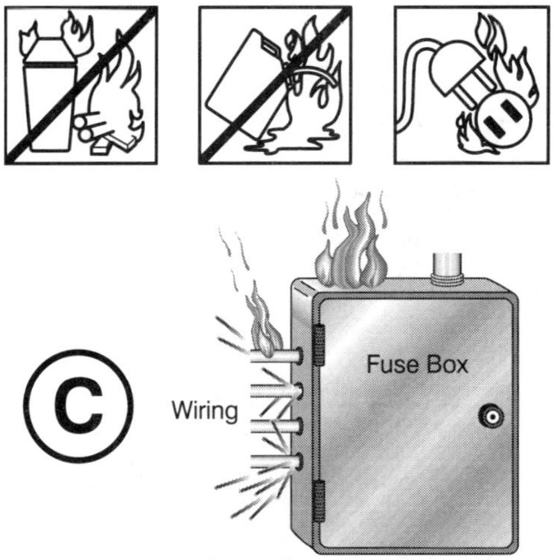

FIGURE 2-3 Type C fire extinguisher. The symbol of a blue circle is for electrical fires.

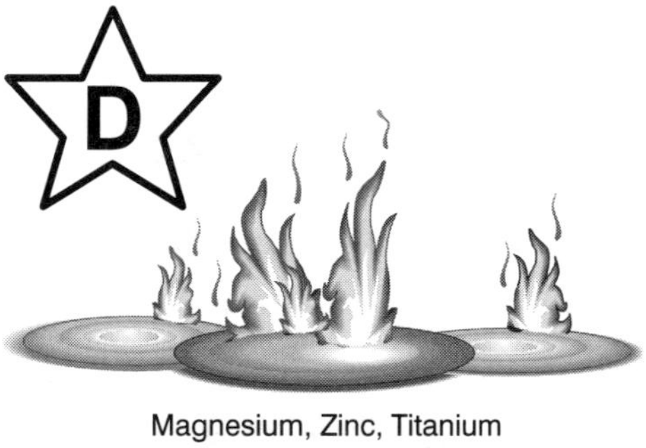

FIGURE 2-4 Type D fire extinguisher. The symbol of a yellow star is for metal fires such as burning magnesium, zinc, and titanium.

radiated by the welding arc can affect the eyes. Contact of the skin to heat parts produced during welding can cause from first-degree to the most severe third-degree burns. Sparks produced from welding will destroy clothing, damage the skin, start **flammable** materials on fire, and cause volatile substances to explode. Electric shock can cause injury and even death (Figure 2-5).

Discussion about the equipment for GMAW and FCAW will center on safety in these five areas of concern. Using the proper welding equipment and supplies is essential for your own protection. (Pay close attention to the suggested safety guidelines listed at the end of this unit.) A welder should become more aware of the dangers presented by smoke, ultraviolet light, heat, sparks, and shock. Always take the necessary safety precautions and remember your ultimate safety depends on you.

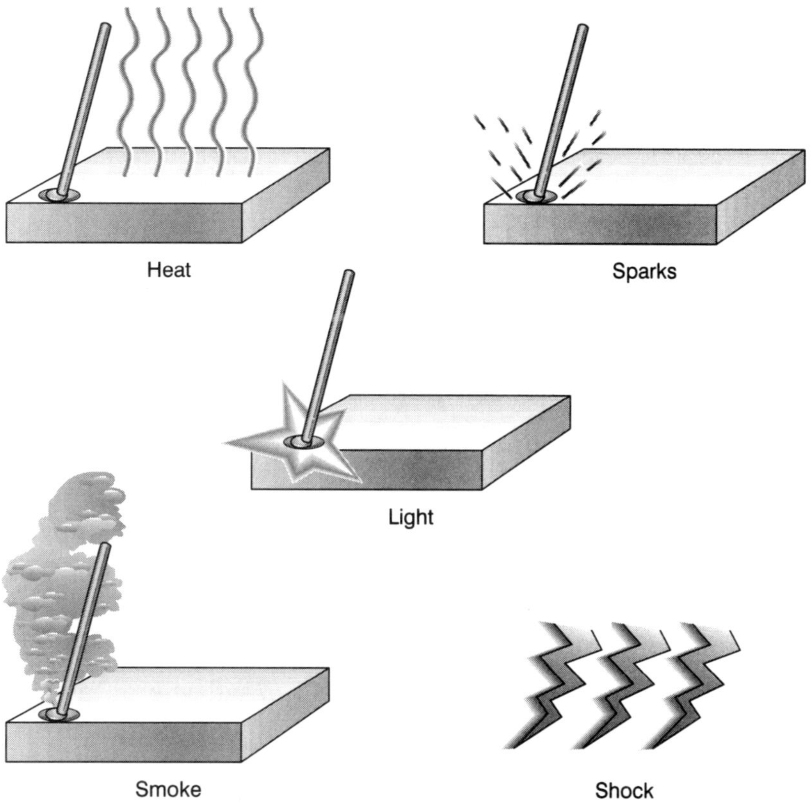

FIGURE 2-5 Five areas of concern associated with GMAW and FCAW.

REVIEW

1. What is the difference between a type A fire extinguisher and a type B fire extinguisher?
2. What is the difference between a type C fire extinguisher and a type D fire extinguisher?

Lab Exercises

1. Identify the types of fire extinguishers in the shop and their recommended uses.
2. Describe the procedure for getting a fire extinguisher in position to put out a fire.

POWER SOURCES

Choice of Welding Machines

The power source is the welding machine for either gas metal or flux cored arc welding. Developments in technology have produced a wide range of power sources from which to choose. Selection of a power source should depend on the requirements of the particular welding process. This section will focus on power sources used for GMAW and FCAW. Some power sources are adaptable for use with either process in addition to welding processes not covered in this book, but the design of the feeder system may restrict application to either gas metal arc welding or flux cored arc welding.

A welder should consider the following information when choosing a power source: repair requirements, portability, safety, purchase price, parts availability, and electrical code requirements.

In addition, the welder should consider the following:

1. Does the welding power source operate on single-phase or three-phase input power? Match the requirements of the welding power source with the input power available on site by the local electric company.
2. Is the welding power source of sufficient amperage (amp) or output current for the job requirements? Not having enough capacity will limit the size of the electrode, which may affect weld quality and increase welding time. Some power sources have rated output less than 100% duty cycle.
3. What type of metal transfer will be used? **Short circuiting transfer? Spray-arc transfer? Globular transfer?** Lower voltage at lower output settings make short circuiting transfer suitable for welding thinner metal. See Unit 4, Metal Transfers, for more detail.

Schools and industrial welding applications use input current provided by three-phase power. Three-phase power is less expensive to operate, but more expensive to install. Conversely, household current for the home workshop generally runs on single-phase input power. Single-phase power sources of 110 volts are becoming popular power sources for gas metal arc welding applications (Figure 2-6). A single-phase power source with a voltage input of 230 remains standard (Figure 2-7). A 230-volt input is the same voltage necessary to operate the kitchen stove.

The selection of a power source by the job requirements should be made partly on the thickness of the base metal. A 1/16-inch **fillet weld** will join two pieces of 1/16-inch-thick sheet metal using a **lap joint** in the flat position with one pass. For example, short circuiting transfer can be chosen, using 0.035-inch diameter **electrode** at settings of 19 to 22 arc volts and 115 to 125 amperes. The selection of a power source within the 150 amperage maximum range should be adequate for welding sheet metal.

A 1/2-inch **V-groove weld** will join two pieces of 1/2-inch-thick steel plate using a **butt joint** in the flat position with three passes. For example, spray-arc transfer can be chosen, using 1/16-inch diameter electrode at settings of 29 to 30 arc volts and 300 to 450 amperes. Or, 1/2-inch V-groove weld will join the same two pieces of 1/2-inch-thick steel plate using a butt joint in the flat position with two passes by flux cored arc welding. In this example, the choice can be a 3/32-inch diameter electrode at settings of 30 arc volts and

FIGURE 2-6 Technology has expanded the portability with power sources that operate at 110 volts and can be plugged in practically anywhere.

FIGURE 2-7 A single-phase 230-volt power source produces higher amperages than 110-volt units.

FIGURE 2-8 The choice of power sources is extensive.

480 amperes. The selection of a three-phase power source above these amperage ranges (from 500 to 650 amperes) should be adequate for welding plate in some procedures.

The choice of a power source will depend on the welding application (Figure 2-8). Regardless of the selection based on home, school, or industrial use, careful thought must be given before making a purchasing decision. A qualified electrician should make wiring changes necessary to accommodate the installation of any power source for welding. The insurance policy should also be examined to be sure of adequate coverage.

The minimum electrical information for rating power sources as determined by the National Electrical Manufacturers Association (NEMA) is stated on a nameplate. This includes NEMA class designation. The nameplate must be attached to the power source (Figure 2-9). This information should be brought to the attention of the electrician. Under no circumstance should a welder remove the outer covering of the power source. Any attention to the internal components of the power source must be left to the service of a qualified electrician.

Duty Cycle

Electrical components heat up under the load of welding. So unless the power source is designed for continuous welding, it will deliver power for a limited time. Each power source is therefore rated with a **duty cycle.** A duty cycle is the recommended percentage of time a power source should be under the load of welding. A 30% duty cycle means the power source is rated for welding 3 minutes out of every 10 at its rated output. Rated out-

1.	Manufacturer's name		
2.	Class 1 60%, 80%, or 100% duty cycle Class II 30%, 40%, or 50% duty cycle Class III 20% duty cyclc		
3.	Input voltage	**4.**	Input amperage
5.	Maximum open circuit voltage (OCV)		
6.	Rated load voltage		
7.	Rated load amperage		
8.	Phase	**9.**	RPM at no load
10.	Duty cycle		
11.	Frequency		

FIGURE 2-9 Information found on the nameplate is useful in giving the capabilities of a power source.

put may be, but is not always, the maximum amperage setting. For welding loads at the maximum and even under the maximum, the duty cycle can be a problem for continuous welding that requires 100% duty cycle. This is especially a problem for power sources that can be used for several welding processes. A power source can be designed for both shielded metal arc welding (SMAW) and gas metal arc welding processes. For the following example, the maximum amperage setting is determined for continuous welding.

> The welding machine is a 300 amperage power source with a 60% duty cycle. A 60% duty cycle means the power source is rated for welding 6 minutes out of every 10 at the maximum amperage setting. The estimated maximum amperage is found by taking 300 multiplied by $(.60)^{1/2}$ which equals 232. An answer of 232 indicates continuous welding at 232 amperes and any setting below 232 amperes.

Overheating will most likely never be a problem if the power source is designed for continuous welding; however, you should be familiar with any information provided by the manufacturer. When using a power source that is not rated for continuous welding, the welder may be required to determine the amperage setting limit requiring 100% duty cycle. In such situations the welder should understand how to make the necessary calculations.

Five Types of Power Sources

Five types of power sources are available for welding operations. They are the transformer, rectifier, inverter, gasoline or diesel engine–driven generator, and the motor generator. Each type is discussed in this section.

The **transformer** is an inexpensive AC power source used for shielded metal arc welding (Figure 2-10). High voltage and low amperage input power are transformed into the low voltage and high amperage output power necessary for welding. The transformer power source is normally single phase. The National Electrical Manufacturers Association (NEMA) Class III transformer power

FIGURE 2-10 The transformer power source is popular for shielded metal arc welding.

FIGURE 2-11 A versatile power source designed for gas metal arc welding, gas tungsten arc welding, shielded metal arc welding, and air carbon arc gouging. (Courtesy of ESAB Welding and Cutting Products, Florence, South Carolina.)

source rated with a 20% duty cycle is popular on farms and in shops where some repair and fabrication welding is required.

A rectifier added to a transformer changes AC to DC (Figure 2-11). Gas metal and flux cored arc welding require direct current for welding. Power sources with rectifiers are available in various sizes with input power that can be single phase or three phase.

The **inverter** is a third type of power source (Figure 2-12). As a new development in technology, some of these lightweight welding power sources can weigh less than 50 pounds. Inverters are also manufactured with the capability to handle either single-phase or three-phase input power in one unit. Some inverters have been designed with the capability of performing other welding processes in addition to GMAW and FCAW.

The gasoline or diesel engine–driven generator is a fourth type of power source (Figure 2-13). This power source can be easily transported with little maintenance for operation in extreme weather without fear of breakdown or concern for the welding environment. Where electricity is not available, an engine-driven

FIGURE 2-12 Inverter power source. The XMT 304 is multiprocess. This power source has the capability for shielded metal arc welding, gas metal arc welding, and gas tungsten arc welding. (Courtesy of Miller Electric Manufacturing Company, Appleton, Wisconsin.)

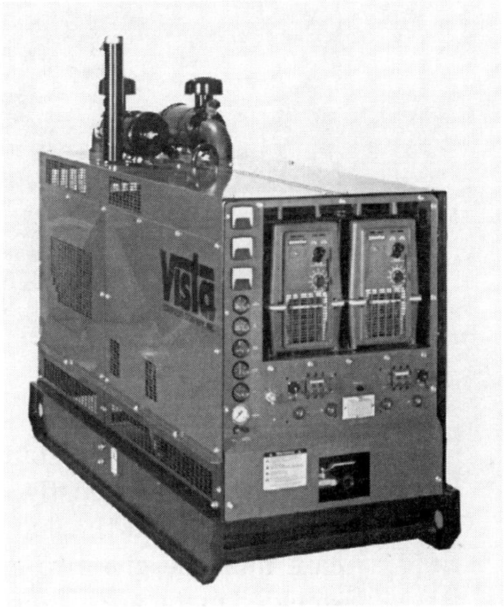

FIGURE 2-13 An engine-driven power source can be hauled by trailer or set on the back of a truck. (Courtesy of Vista Equipment Company, Inc., Crawfordsville, Indiana.)

power source can be dropped into the most remote location to perform the necessary welding. Logging equipment that breaks in a difficult to reach forest can be repaired on the spot by an engine-driven power source.

The **motor generator** is the fifth type of power source (Figure 2-14). The motor generator is similar to the engine-driven power source, but an electric motor rotates an armature to produce the current for welding. The motor generator—noisy and expensive to operate—is the one type of power source becoming a thing of the past as industry incorporates the latest technologies.

The choice of a power source for either GMAW or FCAW can be reduced to its cost. The initial cost of the investment in either equipment is more expensive than shielded metal arc welding. In the long run, GMAW and FCAW become more economical because more welding can be completed in a given time than with any **manual** welding process.

FIGURE 2-14 Motor generator power source.

A welding power source should provide years of service without the worry of costly repair problems, provided the owner is careful to follow the manufacturer's operating instructions. More care must be given to the GMAC or FCAW equipment than to shielded metal arc welding equipment, however, and some critical parts must be replaced periodically.

SAFETY REMINDER

Any piece of electrical equipment used in a shop can cause injury: a grinder, a drill, a sander, or a power source used for welding. Thoughts about safety should be first in your mind at all times. Do not operate shop equipment without wearing both **safety glasses** and **ear plugs.**

Lab Exercises

1. Examine a power source for gas metal arc welding and retrieve the following information:
   a. Manufacturer and brand name
   b. Type of power source

c. Phase (input power)
d. Output current choices
e. Amperage output range

2. Provide the electrical information given on the nameplate.
3. Give the amperage setting for continuous welding or calculate this information from the duty cycle of the given power source.
4. List the welding processes that can be performed in your shop.
5. If the shop has several power sources used for gas metal and flux cored arc welding, describe some differences between them.

EQUIPMENT NECESSARY FOR WELDING

Welding Equipment

The power source, the **feeder system,** the gun, the filler metal, and the shielding gas are the five primary components to GMAW or FCAW (Figure 2-15). Unit 8 will cover filler metals and shielding gases in greater detail. Reasonable care and routine maintenance are required for this equipment to work effectively. Read carefully the operator's handbook supplied by the manufacturer

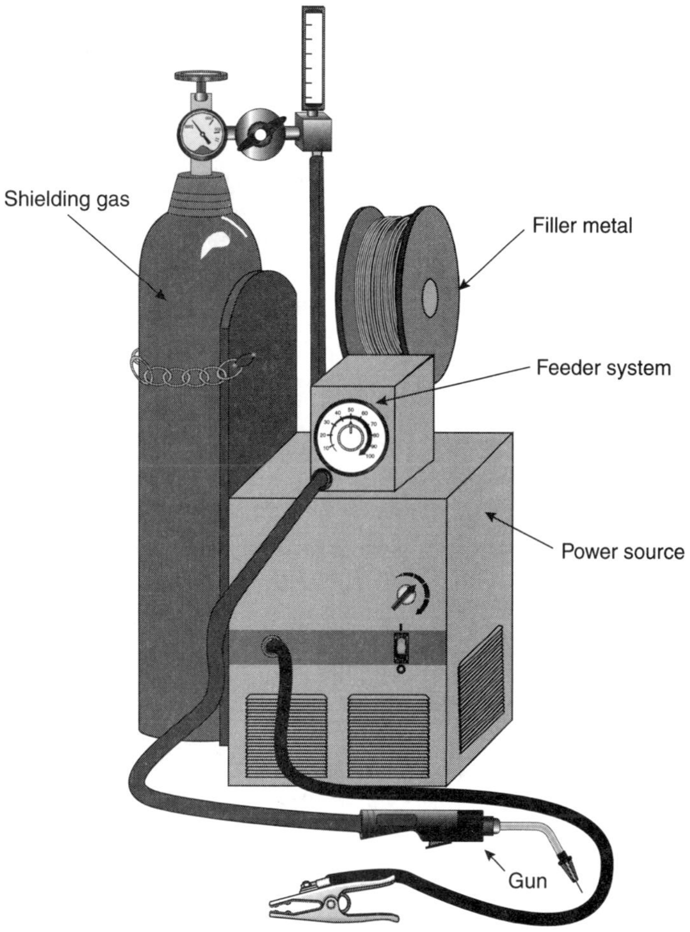

FIGURE 2-15 Primary pieces of welding equipment.

FIGURE 2-16 The control panel provides the welder with the means to do quality work.

to become thoroughly familiar with the particular power source you will be using.

Power Source

The power source contains the main controls (Figure 2-16). Examine the control panel to become familiar with each control knob, switch, button, or keypad. The latest development in technology is the solid-state controls used to preset and program control settings into digital displays. The voltage adjustment and the wire feed speed adjustment are the two major controls located on the power source when all the equipment is self-contained in a single unit.

The voltage adjustment determines the arc length. The wire feed speed adjustment determines the amperage. These two adjustments work in relation to each other. Once the arc voltage is set, adjusting the wire feed speed establishes a range between two extremes. The two undesirable extremes are (1) stubbing the wire on the base metal and (2) melting the wire back onto the contact tube (Figure 2-17).

The **slope** adjustment and **inductance** adjustment are automatically set on most power sources; however, some power sources contain fine adjustment for these controls (Figure 2-18). Slope is the relationship of voltage to amperage that can be plotted as a curve on a graph (Figure 2-19). Slope can be used in combination with the voltage adjustment. Increasing slope causes a decrease in amperage, which can reduce **spatter.** Inductance is

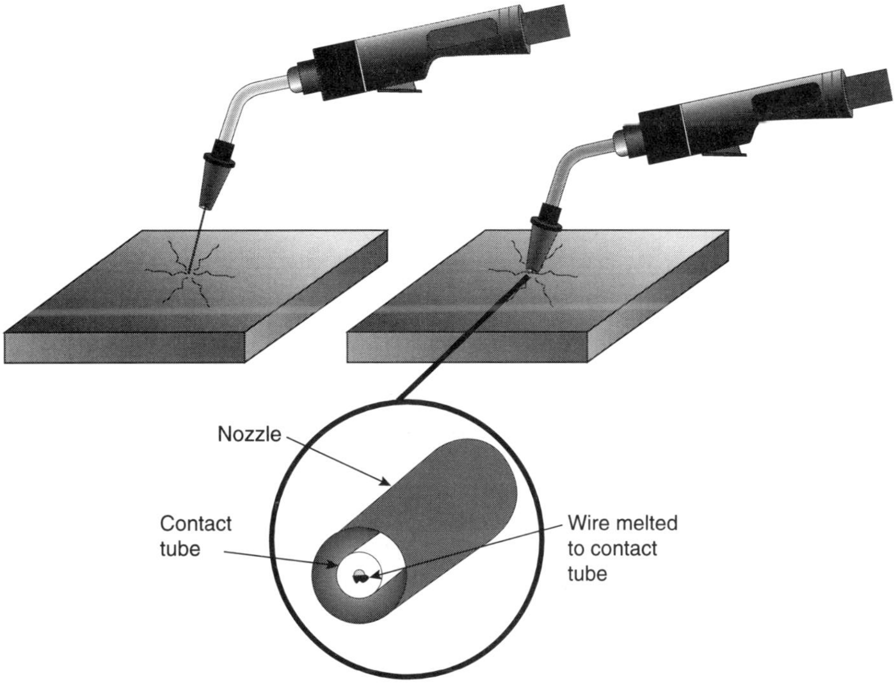

FIGURE 2-17 Stubbing the wire on the base metal and melting the wire onto the contact tip are two fundamental problems encountered when making adjustments for welding.

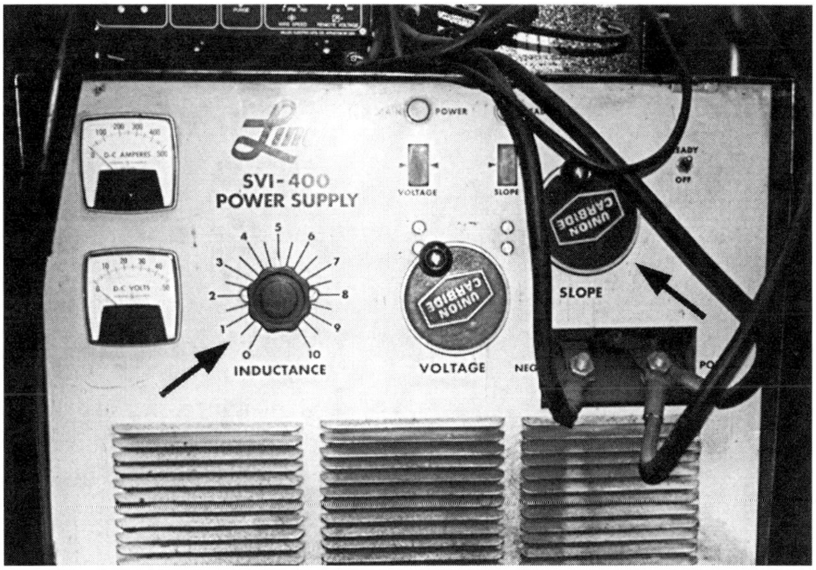

FIGURE 2-18 Slope and inductance give the welder more control.

the characteristic of the electrical circuit that controls the time of amperage rise as the electrode short-circuits (or arcs) during welding. Increasing inductance causes an increase in arc time (wetness of the weld pool) as the amperage rises. Inductance also reduces spatter by controlling the rate of amperage rise. When using any power source with slope and inductance controls, begin by setting these parameters in the middle range.

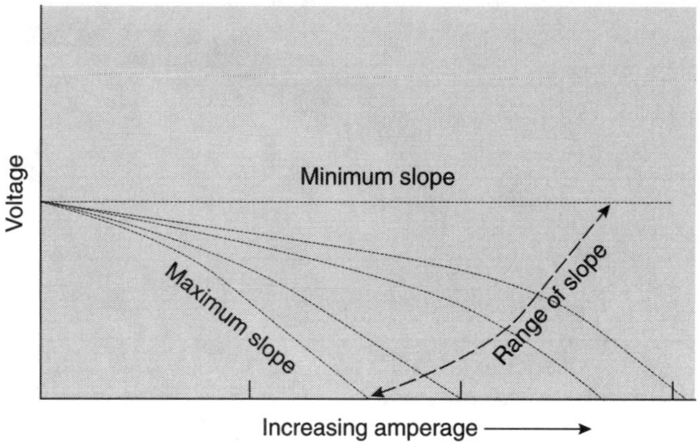

FIGURE 2-19 Increasing slope causes a decreasing amperage which can reduce spatter.

FIGURE 2-20 Feeder system.

Feeder System

The feeder system contains the electric motor with its drive rolls (Figure 2-20). The drive rolls are adjusted or changed according to the size of the wire being used. The drive rolls push the electrode, or wire, through to the gun. On some systems the electrode is threaded through with the tension spring loaded, whereas on other systems pressure is applied to adjustable drive rolls. Care has to be taken to not overtighten the drive rolls so as to distort or flatten the electrode. Overtightening can cause the wire to become tangled in the system, the result of which is called **birdnesting** (Figure 2-21).

On some power sources, the feeder system can be conveniently removed and carried to the work to conveniently make adjustments of the parameters on the spot. Some units are designed with keypad panels for finger touch control. The feeder system contains the wire feed speed control. Another control found on some systems is the mode selection to choose between spot, intermittent, and continuous welding. The gas purge can be used to adjust the shielding gas on the **flowmeter** (Figure 2-22). A jog switch moves the electrode through the feed system without electrical current. The burnback control is used to prevent the wire from sticking in the weld pool. The timer is used to time spot welds. The wire reel braking system stops the rotating spool of wire when the welding stops.

FIGURE 2-21 Adjust drive rolls properly to avoid birdnesting.

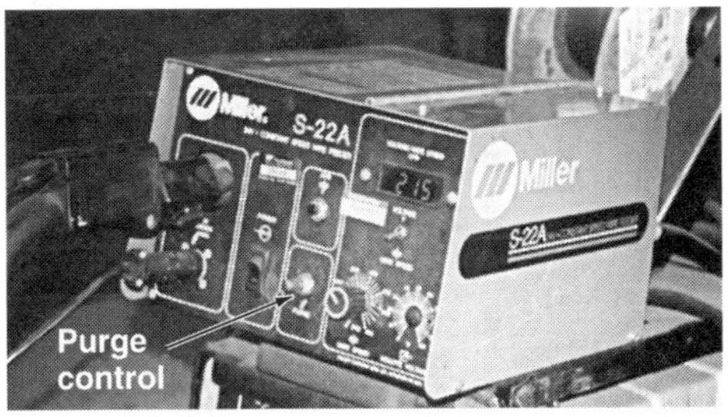

FIGURE 2-22 Gas purge is available on some feeder systems.

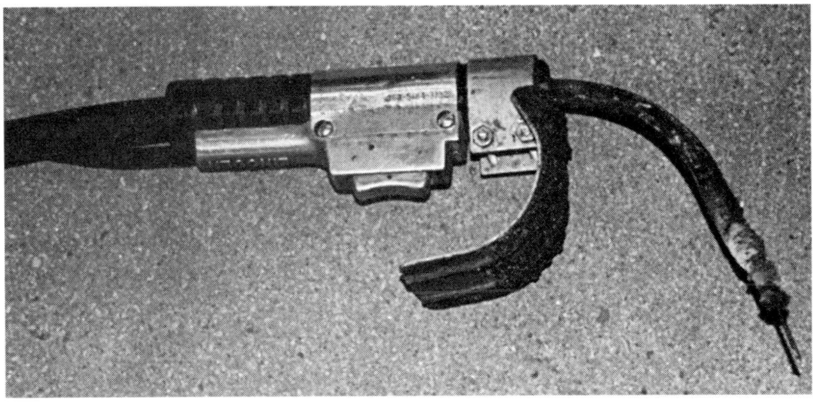

FIGURE 2-23 Gun has on/off trigger switch.

Gun

Many different styles of guns are available. Guns can be of differing weights and are designed with varying gun angles which are either permanent or adjustable. Most guns, however, contain the following components.

The gun or torch has a trigger, or on/off switch, that energizes the gun with electrical current and activates the feeder system (Figure 2-23). The feeder system either pushes or pulls the wire

FIGURE 2-24 Spool gun permits welder mobility away from the power source.

through the gun. Some systems use a combination of these methods. When the drive rolls are located in the feeder system, the wire is pushed through the gun. When the drive rolls are located in the gun, the wire is pulled through the gun. With a **spool gun,** the drive rolls and a small, 1 to 2 pound spool of wire are located in the gun (Figure 2-24). A spool gun is used when the welding covers a large area of the work space and the power source cannot be brought close enough to the base metal. A spool gun is also used with a soft wire such as **aluminum,** which can kink or twist when moving through the gun.

The gun consists of a **nozzle** to direct the gas flow and an **insulator** to keep the nozzle from fusing to the base metal (Figure 2--25). Electrical contact should only occur at the point of the contact tube, where the wire passes through. The gun also contains a **liner,** which carries the wire from the spool; the liner should be matched to the size of the electrode, or filler metal, being used. The liner is connected to a **gas diffuser,** which is connected to the liner by a locking screw. Shielding gas flows out through the gas diffuser. The contact tube screws into the gas diffuser, and the gas diffuser screws into the **conductor tube,** which transmits electrical current through from the copper cable. The power cables are available in standard lengths of 10, 12, 15, and 25 feet. The conductor tube

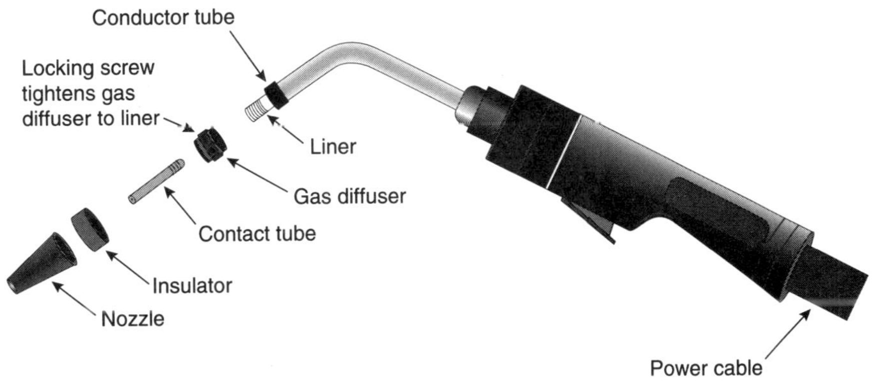

FIGURE 2-25 Parts of a gun.

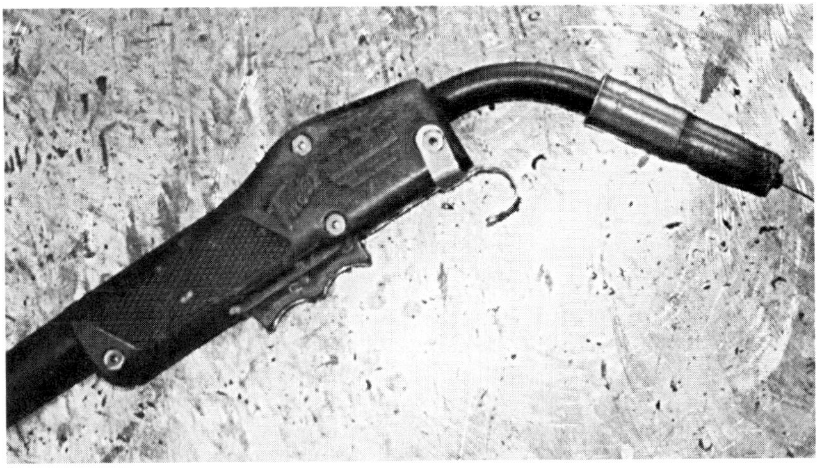

FIGURE 2-26 Air cooled gun.

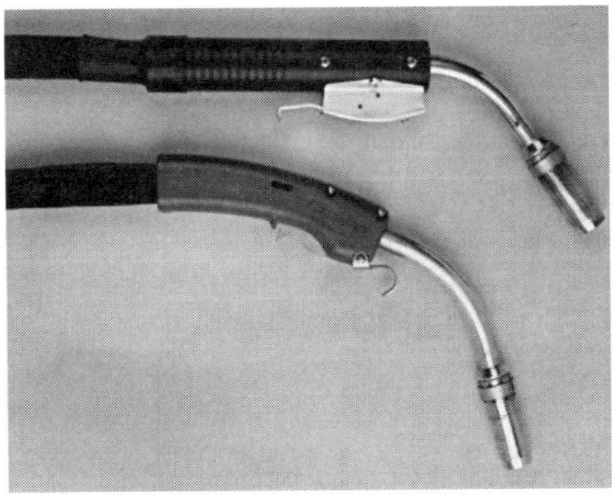

FIGURE 2-27 Water-cooled guns. (Courtesy of Bernard Welding Equipment Company, Beecher, Illinois.)

and the power cable are insulated; the handle fits over the conductor tube and the welder holds the power cable with a gloved hand.

The guns are rated for duty cycle. Guns with a lower rating are air cooled and guns with a higher rating are water cooled (Figure 2-26). The air cooled guns are usually heavier than water-cooled guns with the same rating (Figure 2-27). The water-cooled guns are designed for larger diameter wires at higher amperages, and can be connected to municipal water which requires a filter and a regulator

FIGURE 2-28 Water circulating system.

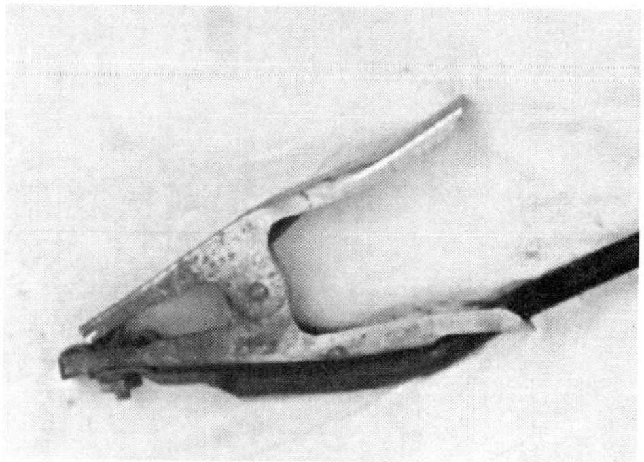

FIGURE 2-29 Workpiece connection with lead.

to meet the water pressure recommended by the manufacturer. The water-cooled guns can also be connected to a closed system, using a water circulating system that consists of a reservoir, pump, and fan-cooled radiator to dissipate the heat (Figure 2-28).

Workpiece Connection

There are many different styles of workpiece connections. The workpiece connection consists of a clamp that completes the electrical circuit necessary for welding (Figure 2-29). The workpiece connection is attached to the power source by a **lead** (cable). The clamp must be secured to the base metal for current to flow as required for welding. The lead should be fitted tightly to the clamp for current to flow properly.

If there is a problem striking an arc, the workpiece connection should be examined immediately. A bad connection because of **rust,** scale, or paint on the base metal can prevent completing an electric circuit. The cause can also be a loose fitting connection at a point where the **workpiece lead** is attached to the workpiece connection. The workpiece connection should always be attached as close to the weld as possible. If a connection is not made directly to the base metal at the point of welding, undesirable arcing may occur which can damage another part (Figure 2-30).

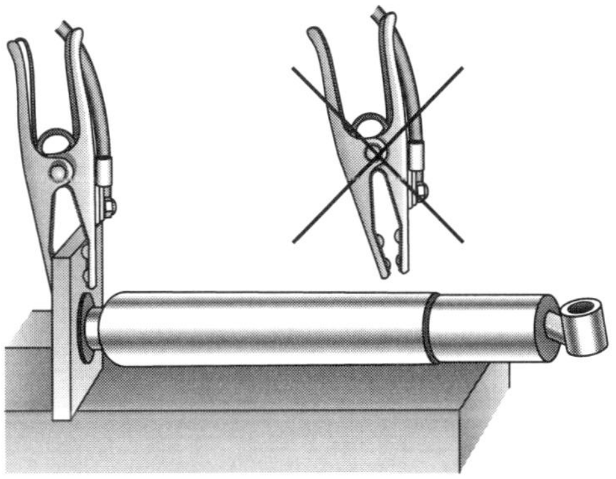

FIGURE 2-30 Placement of the workpiece connection can be important to avoid damage to equipment. In this example it is important to keep the current from passing through the hydraulic cylinder and thus arcing to avoid damaging the cylinder wall.

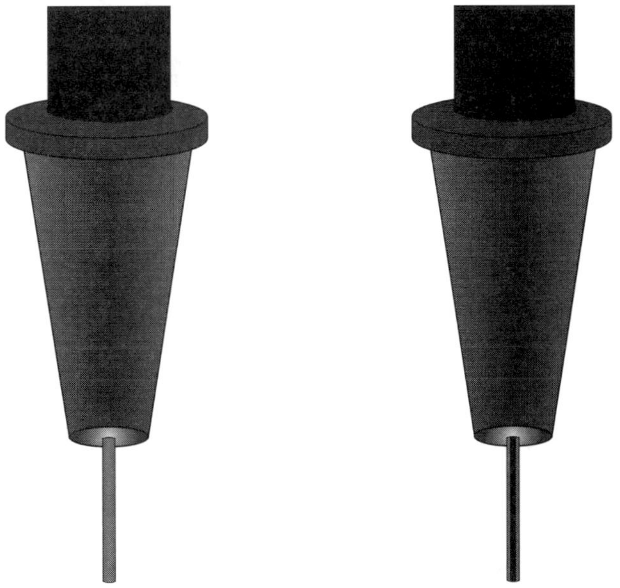

FIGURE 2-31 The solid wire has no flux and the tubular wire contains flux inside the wire.

SAFETY REMINDER

Special care must always be taken in placing the workpiece connection, because electricity will find the path of least resistance. Know the path the electricity will travel to reach the point of welding.

Filler Metal

The filler metal, electrode, or wire is the material added to the joint by welding. The filler metal melts, mixing with the base metal to form the weld that becomes a permanent part of the joint. The filler metal is designed by the manufacturer according to **specifications** subject to testing, inspection, and certification. The filler metal can be a solid or a tubular wire containing flux (Figure 2-31). By affixing to its product the **American**

TABLE 2-1 AWS specifications for steel electrodes

Base Metal	*AWS Specification*
Carbon steel	A5.18 / GMAW
Carbon steel	A5.20 / FCAW
Low-alloy steel	A5.28 / GMAW
Low-alloy steel	A5.29 / FCAW
Stainless steels	A5.9 / GMAW
Stainless steels	A5.22 / FCAW

Welding Society (AWS) specifications with classifications, the manufacturer certifies that its products meet the requirements of the AWS under code A5 for GMAW and FCAW (Table 2-1). The various wires are discussed in Unit 8.

Shielding Gas

The shielding gas serves several functions, specifically it (1) protects the weld pool from contamination by the air, (2) stabilizes the arc, (3) transfers heat, (4) helps control the appearance of the weld bead, and (5) aids root penetration. The shielding gas system consists basically of a gas cylinder and **regulator**/flowmeter, connected to the feeder system. The major shielding gas(es), depending on the welding, can be the **inert gases helium** and **argon,** or the **active gas carbon dioxide.** These gases can be used in combination with each other or by the addition in a percentage of oxygen.

The shielding gas can be supplied by individual cylinders or furnished economically in bulk through a manifold. Gas cylinders are available in several sizes (Figure 2-32). Gas provided by an individual cylinder has its pressure controlled by use of a regulator, or a regulator/flowmeter (Figure 2-33). A pressure regulator is used to set the pressure of the shielding gas flowing out of the gun. A flowmeter added to a regulator adjusts the flow, measured in cubic feet per hour (cfh). With a flowmeter, volume is indicated on a scale by a ball riding on top of the flow inside a clear tube. Flowmeters can be calibrated for one particular gas or a combination of several gases. With a regulator/flowmeter combination, the pressure from the regulator to the flowmeter is preset by the manufacturer.

SAFETY REMINDER

Never use petroleum-based products near pressurized cylinders, nor operate tools containing petroleum-based products on high-pressure fittings. Never do any welding near a pressurized cylinder to avoid an accidental arc strike on the cylinder wall.

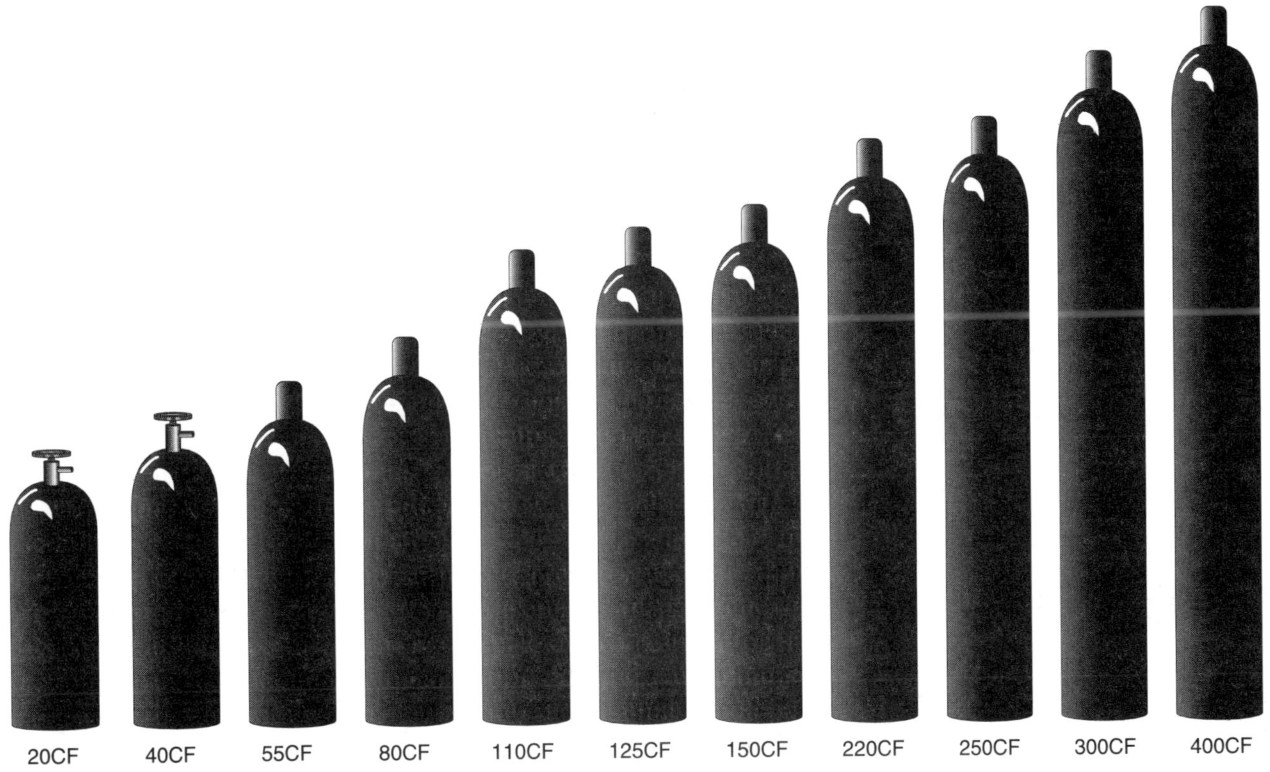

FIGURE 2-32 Gas cylinders are available in many sizes.

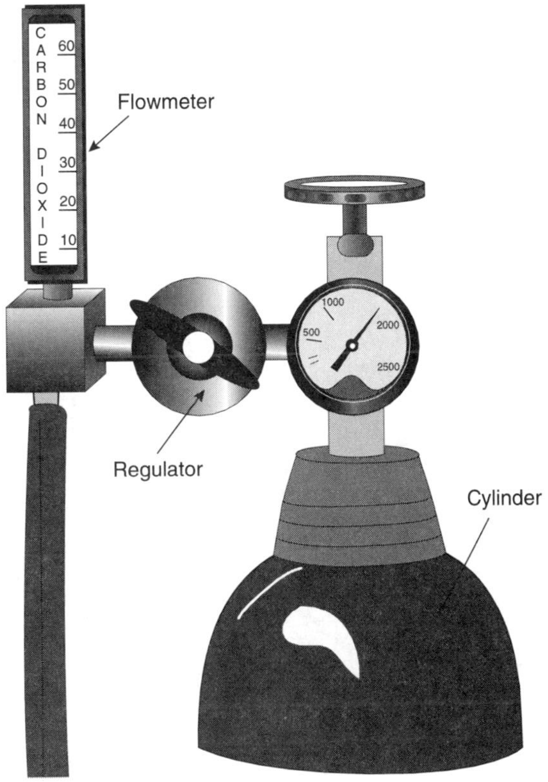

FIGURE 2-33 Regulator/flowmeter dispenses gas in a controlled manner.

Lab Exercises

1. Locate the voltage adjustment and wire speed adjustment on two or three power sources.
2. Are these two or three power sources set for GMAW and FCAW? How do you know?
3. Are any water-cooled guns located in the shop? How do you know?
4. What size wire is being used on two different power sources?
5. Disassemble a gun and label its parts.
6. List the shielding gases used in the shop.

SAFETY EQUIPMENT

Safety Glasses

> **ARC FLASH**
>
> Have your eyes checked, if you haven't done so recently. Welding requires good vision to do quality work.

No shop work should be performed without wearing safety glasses (Figure 2-34). Safety glasses are insufficient without side shields. Besides offering protection from normal shop activities, safety glasses protect the eyes from slag (a nonmetal by-product of the weld pool produced during flux cored arc welding). Protect the most precious of your five senses, and never wear safety glasses without a welding helmet and the required filter lens to protect against **arc flash.**

Ear Plugs

Ear plugs are probably the most overlooked piece of safety equipment (Figure 2-35). Ear plugs may be neglected because hearing loss develops gradually. Wearing the appropriate ear plugs helps protect the ears from damaging, high-frequency sound waves produced by grinding. Under the decibel scale for measuring sound, the threshold of pain is 140 decibels. Any sustained exposure from between 120 to 140 decibels can cause hearing loss.

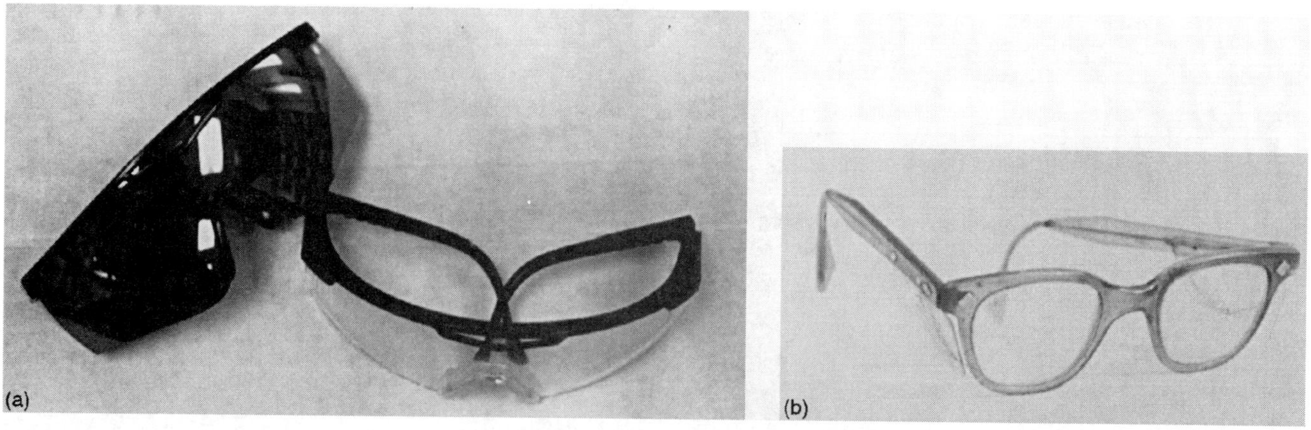

FIGURE 2-34 (Photo) Safety glasses with side shields come in many styles, but all must protect the eyes from hazards threatening from both the front and the sides.

Ear plugs also prevent hot slag and sparks from entering the ear. Do not take this small piece of safety equipment for granted. Continue to enjoy the beauty of music and those wonderful sounds of nature.

Respirators

The respirator is required whenever there is a need to reduce exposure to respiratory contaminants potentially harmful to the body. The base metal, its different types of coatings, electrodes, and objects in the welding area can produce fumes and gases in the smoke given off by welding. The filter and cartridge respirators are one type of air-purifying respirator (Figure 2-36). The air-purifying respirator stops airborne contaminants from passing through the nose and the mouth. Respirators are available in hardware stores.

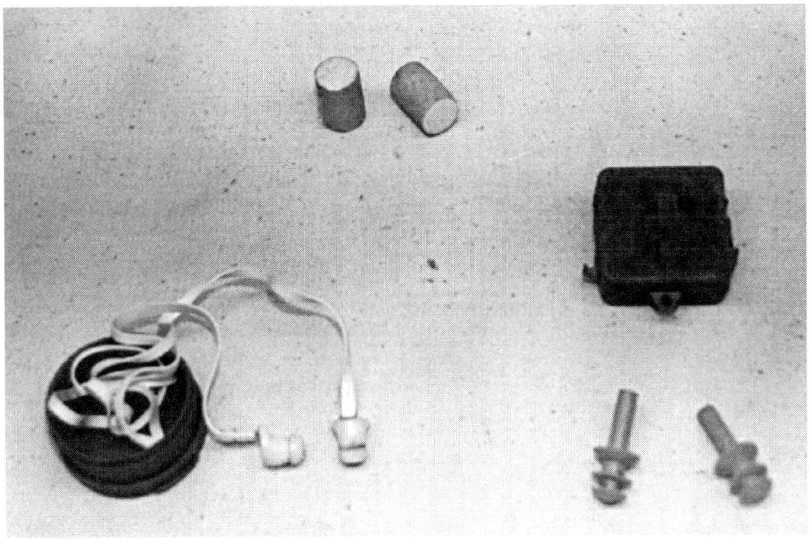

FIGURE 2-35 Ear plugs protect hearing. Choose a pair from the various designs available that are comfortable to wear—and wear them.

FIGURE 2-36 A respirator must be worn in certain environments in which there is danger of inhaling toxic materials.

FIGURE 2-37 Welding helmet with flip-up lens. This type allows the welder to examine the work without having to flip up the entire helmet.

The atmosphere-supplying respirator is the second type of respirator. This respirator is necessary when fumes and gases produced by welding cannot be inhaled, or the welding is being performed in a confined area. If there is ever a question as to the danger from breathing the smoke produced, do not perform the welding without the required protection.

SAFETY REMINDER

Safe practices require that the welder always try to keep the head out of the smoke stream during welding and use a respirator when required.

Welding Helmet

The welding helmet protects the head, face, and eyes from the effects of the arc during welding. Many styles of helmets are available on the market (Figure 2-37). A lightweight helmet is best to avoid neck strain. A helmet with a flip-up filter lens (2 inch by 4½ inch) offers better accuracy in placement of the electrode before critical welding. However, a helmet designed with a larger filter lens (5¼ inch by 4½ inch) which does not flip up increases the field of vision. Some rather expensive helmets convert the lens to the necessary filter instantly upon making an **arc strike** (Figure 2-38).

Always examine the helmet and the filter lens in the lights before welding. A crack in the helmet or the filter lens indicates immediate replacement. The filter lens should meet the shade number for the amperage setting to be used during welding, as shown in Table 2-2. The filter lens is designed to keep the **infrared** and ultraviolet light produced by welding away from the eyes. Follow the lens manufacturer's recommendations to match the filter lens with the welding process. Use as dark a lens, not lighter, as possible to observe both the weld pool and the joint at the point of welding.

FIGURE 2-38 Welding helmet in which light window darkens instantly upon arc strike. This type of helmet is expensive.

TABLE 2-2 Suggested filter lens chart for lenses made from DESAG welding protection glass protection against ultraviolet and infrared radiation (exceeds ANSI Z87.1)

Shade Number	*Amperage Setting*
10	80 to 99 amperes
11	100 to 174 amperes
12	175 to 299 amperes
13	300 to 499 amperes
14	500 to 560 amperes

Be sure to examine the entire filter lens assembly before welding. While checking for cracks, clean the assembly of smoke stains. For some style helmets the filter lens is sandwiched between two clear cover plates. The cover plate nearest the eyes should be shatter resistant. A fiber gasket separates the cover plate on the arc side of the filter lens (Figure 2-39). The filter lens should always be installed according to the manufacturer's recommendations.

Arc flash is a painful and usually temporary eye condition caused by the light of the welding arc. Never weld without a helmet and protect your eyes from the welding arcs of others who may be working nearby. A flash curtain or some type of barrier should be used to screen onlookers.

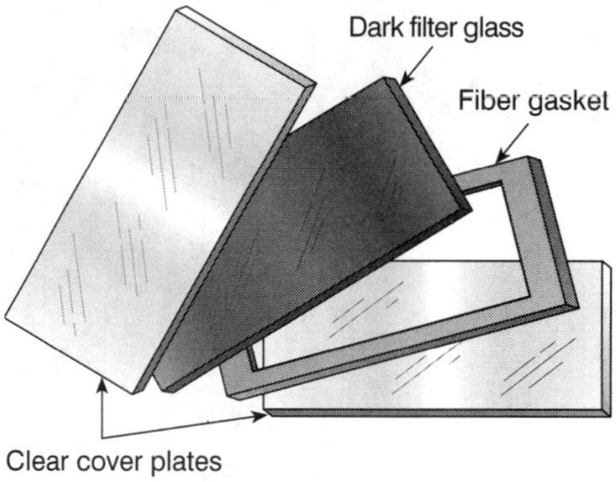

FIGURE 2-39 The entire filter lens assembly should be disassembled and kept clean.

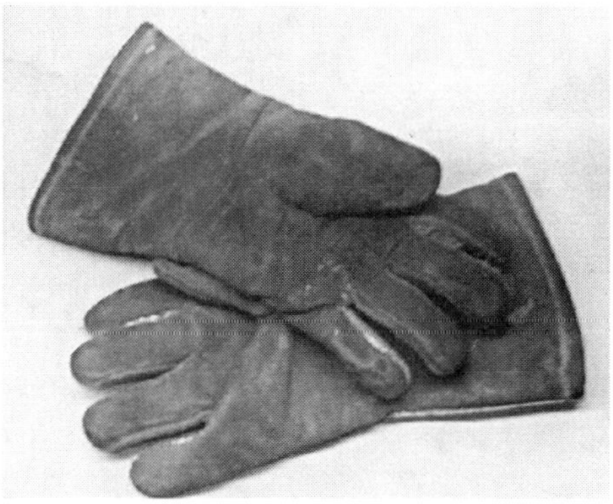

FIGURE 2-40 Welding gloves with gauntlet protection should be worn.

SAFETY REMINDER

As a common courtesy, alert anyone in the immediate area before making an arc strike so eyes can be protected from the flash of the arc.

Welding Gloves

Welding gloves protect the hands and lower arms from burns and electric shock. Only gloves made for welding do an adequate job of handling heat without losing their shape. Gloves with gauntlets that protect the wrists and forearms should be worn (Figure 2-40). Use tongs to handle hot metal; the welding gloves will last longer and the leather will not become stiff (Figure 2-41).

Clothing

Wear leathers, cotton, or even wool, but not synthetic fibers that may burn or melt to the skin. Fire retardant fabrics can also be worn. Clothing should cover the skin to shield the body from the arc, which is more intense than sunlight. Ultraviolet light will burn the skin as will the spatter that welding produces. A leather jacket, leather sleeves, and chaps—although expensive—are preferred. Leather can save clothing from the sparks and spatter of welding, and leather does a better job of protecting the body from burns than either cotton or wool (Figure 2-42).

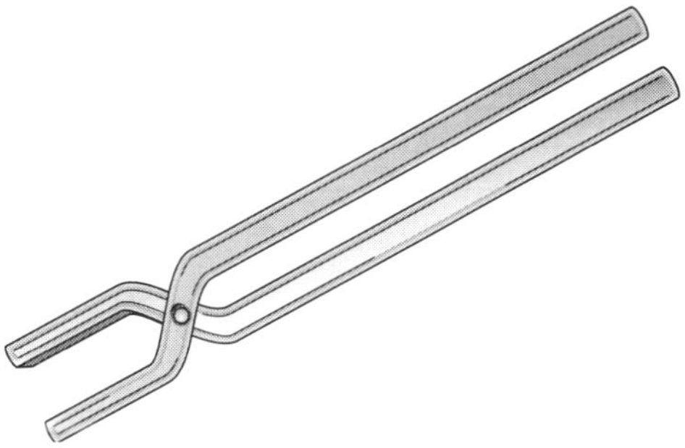

FIGURE 2-41 Using tongs to handle hot metal will extend the life of welding gloves.

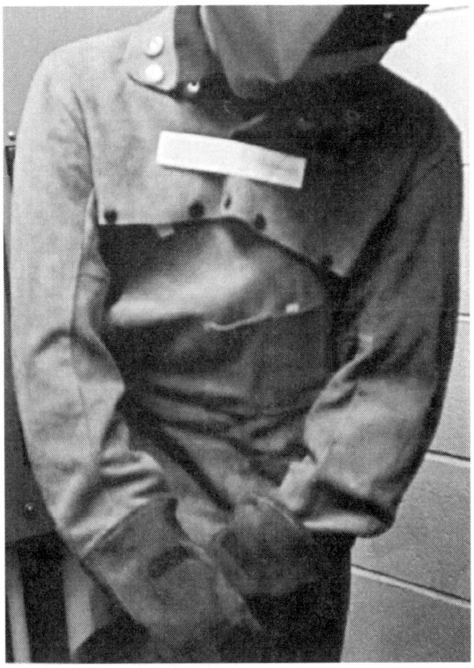

FIGURE 2-42 Leathers protect the clothing and the skin underneath.

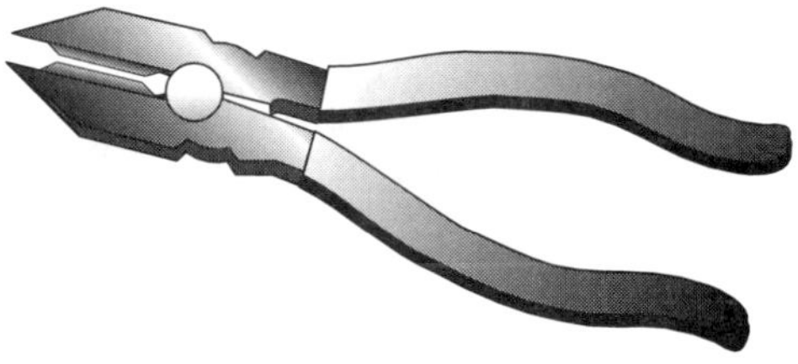

FIGURE 2-43 A side cutter capable of grabbing the wire is handy.

Cap

A cap worn under the welding helmet can protect the scalp from the sparks and spatter of welding. A cap also keeps the smoke and dirt in the welding environment from getting into the hair. The bill of the cap to the rear gives protection to the back of the neck.

Tools

A side cutter is required for cutting the wire filler metals used in GMAW and FCAW (Figure 2-43). A side cutter designed for heavy-duty work will cut easily through various gauges of wire.

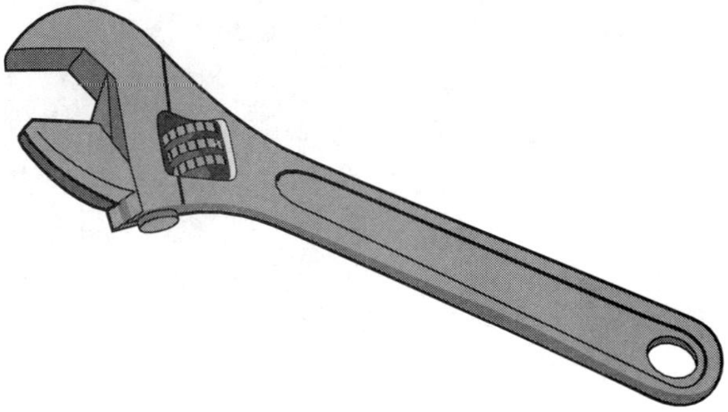

FIGURE 2-44 An adjustable wrench is necessary for gas fittings.

FIGURE 2-45 The slag hammer is a tool of the trade.

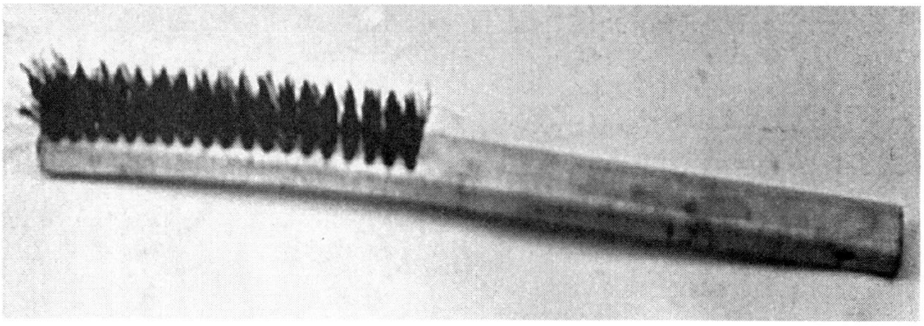

FIGURE 2-46 The wire brush is an essential tool.

An adjustable wrench is necessary for both tightening and loosening the fittings that attach the shielding gas (Figure 2-44). Proper use of this wrench is required once the brass fittings are first tightened by hand to avoid excessive pressure should the threads be misaligned.

Flux cored arc welding has two indispensable tools for doing quality work. The slag hammer is used to chip away the slag that

FIGURE 2-47 Ventilation systems remove the smoke and the fumes of welding. The Torit is a portable system manufactured by the Donaldson Company of Minneapolis, Minnesota.

forms on the weld bead (Figure 2-45). The wire brush is used to remove the slag, smoke stains, and bits of debris (Figure 2-46). Slag remaining on the weld can be trapped inside other weld beads that follow. Slag trapped inside as weld beads are **layered** on top of each other and affect the quality of the weld.

Ventilation

Smoke and the fumes produced during welding, especially FCAW, can be eliminated through a ventilation system. Ventilation systems with specially designed filters are available commercially (Figure 2-47). Such potentially harmful substances as paint, solvents, oils, and galvanized coatings should be removed from the joint surface of the base metal before welding. If this procedure cannot be accomplished, special care must be taken to avoid breathing toxic fumes. Care must always be taken to protect your respiratory system from inhaling the injurious by-products of welding.

SAFETY REMINDERS

Following is a list of safety reminders for GMAW and FCAW.

1. Wear safety glasses with side shields, designed to protect the eyes, for all shop activities.
2. Wear ear plugs in the shop area for protection from high-frequency sound waves.
3. Wear welding gloves to avoid shocks. Gloves should have gauntlets to protect hands and lower arms from burns.
4. Cotton clothing, fire retardant fabrics, and leathers offer protection from exposure to the effects of the arc.
5. Be sure the shop is clear of volatile and flammable material (Figure 2-48).
6. Do not bring pressurized cigarette lighters into the shop.
7. Always wear a welding helmet with the correct filter lens to protect against arc flash.
8. Make sure the welding area is equipped with the appropriate fire extinguisher.
9. Electricity will kill; dry clothing, gloves, and a dry workplace help prevent shocks.
10. Stretch out the workpiece lead and cable attached to the gun; never wrap them around the body.
11. Never weld on pressurized cylinders or on containers that have held volatile substances. Never place welding leads near these cylinders or containers.
12. Make sure any loose connection on the workpiece connection is repaired to maintain electric contact.
13. Keep track of wire cut from the spool. Cut wire can be a shop hazard as it is sharp enough to puncture the skin.
14. Turn off the power source if left unattended.
15. Replace parts of the gun when they wear out.
16. Secure the gas cylinder when setting up for welding and take the necessary precautions to avoid an arc strike on a gas cylinder (Figure 2-49).
17. Be sure the welding area is properly ventilated.
18. Use common sense and pay close attention to the work. Avoid daydreaming and distractions.

FIGURE 2-48 Remove volatile and flammable materials from the welding area. Gasoline, paint thinner, and oily rags can be dangerous anywhere, but especially in the welding area.

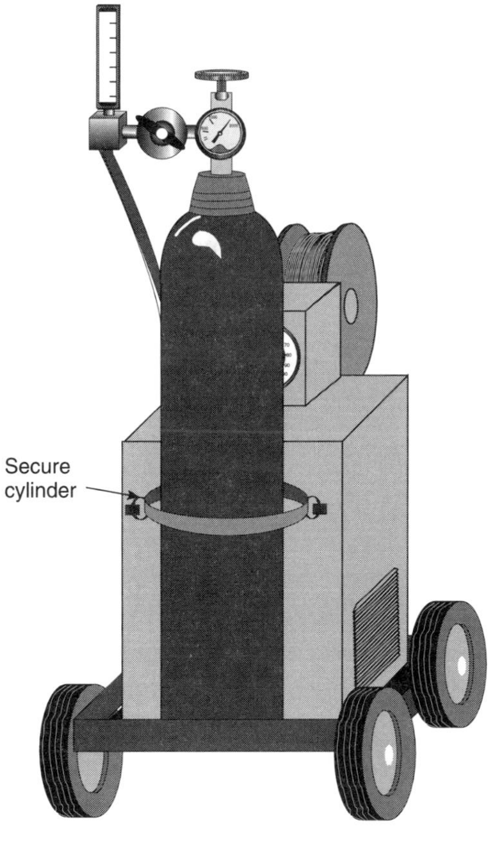

FIGURE 2-49 The cylinder must be secured to protect an accidental fall.

Lab Exercises

1. Develop a safety program or become familiar with the safety program being used in your shop. For example, is there someone responsible for handling hazardous situations such as tripping hazards, electrical hazards, or dangerous work situations? Officers can be chosen, with a description of each job such as a first aid officer and a safety officer.

2. Does the shop have a first aid kit? If not, how will minor injuries be handled?

3. What is the procedure the shop will follow in case of a major injury?

4. Become familiar with the welding helmet you will be using.
   a. Adjust it to ensure proper fit on your head.
   b. Examine for cracks in the helmet and the filter lens.
   c. Clean the entire filter lens assembly of smoke stains.
   d. What is the shade number of the filter lens?
   e. Is the shade number sufficient for the welding you will be doing?

REVIEW QUESTIONS FOR UNIT 2

Multiple Choice

Choose the *best* answer to complete the statement.

1. An A-B-C fire extinguisher is for use on:
   a. Electrical fires.
   b. Metal fires.
   c. Combustibles.
   d. Answers a. and c. above.
   e. All of the above.

2. Smoke, light, heat, ______, and ______ are five areas of concern in welding.
   a. Safety and injury
   b. Stress and cold
   c. Sparks and shock
   d. Gasoline and fire

3. Information to consider in choosing a power source includes:
   a. Repair requirements.
   b. Portability.
   c. Safety.
   d. Noise.
   e. All of the above.
   f. Answers a., b., and c. above.

4. Nameplate information on the power source includes:
   a. Manufacturer's name.
   b. Duty cycle.
   c. Class.
   d. Rated load amperage.
   e. All of the above.

5. ______ means the power source is rated for 4 minutes out of every 10.
   a. 20% duty cycle
   b. 40% duty cycle

c. 60% duty cycle
d. 10% duty cycle
e. None of the above

6. Transformer power source:
   a. Is used for gas metal arc welding.
   b. Is popular on farms.
   c. Converts high voltage and low amperage to low voltage and high amperage.
   d. Is all of the above.
   e. Is b. and c. above.
7. Common shielding gas:
   a. Is argon.
   b. Is carbon dioxide.
   c. Is acetylene.
   d. Is nitrogen.
   e. Includes a. and b. above.
8. ______ determines arc length.
   a. Feeder system
   b. Voltage adjustment
   c. Wire feed speed
   d. Inductance
   e. None of the above.
9. The point where electrical energy passes to the wire is the:
   a. Nozzle.
   b. Liner.
   c. Diffuser.
   d. Conductor tube.
   e. Contact tube.

Short Answer

1. Discuss the functions of the shielding gas.
2. State the purpose of wearing safety glasses.
3. Explain the difference between an air-purifying respirator and an atmosphere-supplying respirator.
4. Give the shade number for welding 100 to 174 amperes.
5. Explain what a flowmeter measures.
6. Define the term *inert gas* and provide two examples.
7. Explain the purpose of a gas regulator.
8. Explain why leathers are preferred to other materials when welding.
9. Explain what a ventilation system does.

ADDITIONAL ACTIVITIES

Fill-In the Blanks

1. Never weld on ______________ cylinders.
2. ______________ keeps the nozzle from fusing to the base metal.
3. ______________ ______________ consists of a clamp and a lead attached to the power source.

4. ____________ ____________ guns are designed for larger diameter wires.
5. Shielding gas flows out through the gas ____________.
6. ____________ ____________ can be solid or tubular.
7. ____________ ____________ push or pull the wire, and must be changed according to the size of the wire.
8. ____________ carries wire from the spool to the gun.
9. ____________ ____________ used for welding that covers a large area.
10. ____________ substances include oils, paints, and solvents.

Writing Skills

1. Write a paragraph using the following terms: *ultraviolet, skin, heat, eyes, flammable,* and *fire.*
2. Write a paragraph to explain what happens when the trigger on the gun is squeezed when the power is turned on.
3. Close the text and from memory write a list of all the safety points listed in this unit.

Math

A regulator/flowmeter is required to control the pressure and regulate the gas coming out of a cylinder. Pressure is measured in pounds per square inch; however, flow is measured in cubic feet per hour.

To square any number, multiply the number by itself.

Examples: $1^2 = 1\ (1 \times 1)$; $2^2 = 4\ (2 \times 2)$; $3^2 = 9\ (3 \times 3)$.

1. $5^2 = ?$ **2.** $9^2 = ?$ **3.** $15^2 = ?$ **4.** $40^2 = ?$ **5.** $45^2 = ?$

Surface area can be measured in square inches, feet, or yards.

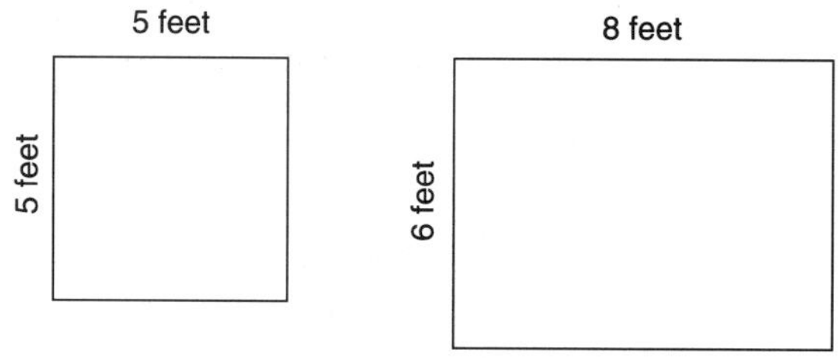

Examples: $5 \times 5 = 25$ square feet; $8 \times 6 = 48$ square feet.

Find the surface area measurements for the following squares.

6.

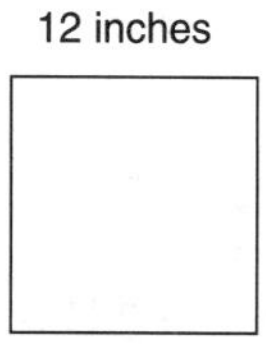

7.

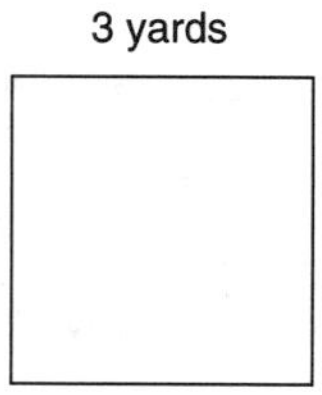

8. 

9. 

Find the surface area measurements for the following rectangles.

10. 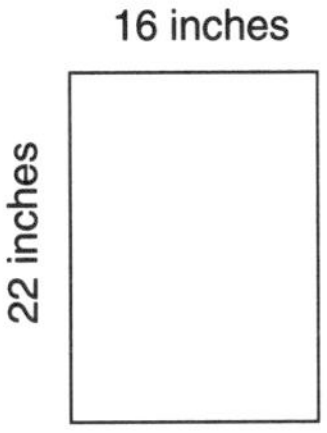

11. 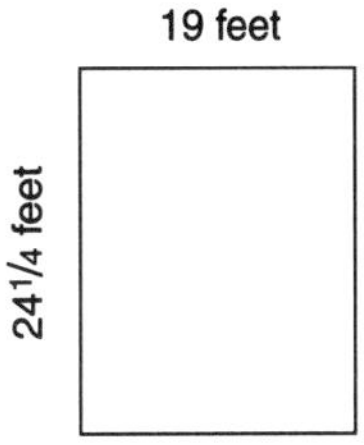

12. 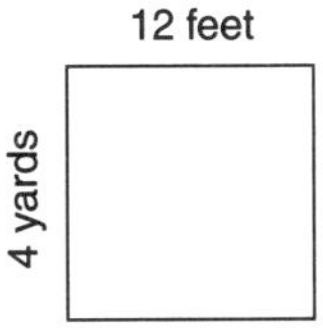

13. 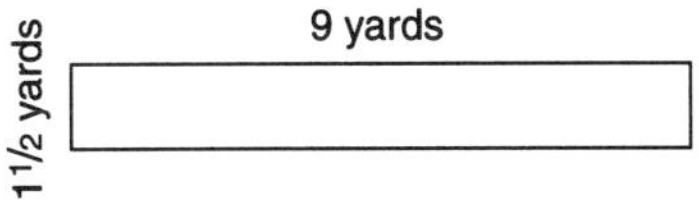

Volume is measured in length by width by height (L × W × H).

Example: Container is 10 feet long, 5 feet wide, and 3 feet high.
$10 \times 5 \times 3 = 150$ cubic feet.

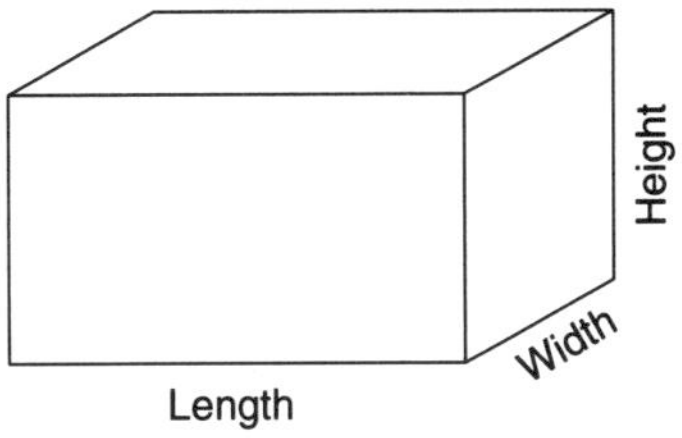

14. 3 feet × 2 feet × 1 foot = ?

15. 20 inches × 9 inches × 4 inches = ?

16. 4 feet × 1 foot 4 inches × 8 inches = ?

17. 6 feet × 6 feet × 6 feet = ?

18. 10 feet^3 = ?

A cylinder's pressure can be 2,200 pounds per square inch of surface area at 70° Fahrenheit, yet the volume of gas available for welding depends on the size of the cylinder. This volume is measured in cubic feet.

A full K cylinder contains 242 cubic feet of argon.

19. How long will the cylinder last at a flow rate of 20 cubic feet per hour?

20. How long will the cylinder last at a flow rate of 30 cubic feet per hour?

A full HK cylinder contains 213 cubic feet of helium.

21. How long will thc cylindcr last at a flow rate of 25 cubic feet per hour?

22. How long will the cylinder last at a flow rate of 30 cubic feet per hour?

Puzzlers

1. Which type of power source would you choose that is both mobile and heavy duty enough to be capable of air carbon arc cutting, and why?

2. You are the first person to arrive on the scene of a serious accident in the shop. What should you do?

UNIT 3

Welding Materials

METALS FOR WELDING

Properties of Metal

Most metals can be joined by welding. Metal has many **structural shapes** with unlimited possibilities for welding ideas (Figure 3-1). Its **properties** are the characteristics by which a metal is identified. Metal has many properties, for example, it can be light or heavy, hard or soft. Metals can also be **ferrous** or **nonferrous.** Ferrous metals are primarily iron and include the carbon steels, **alloy steels,** and **cast irons.** Nonferrous metals, which are nonmagnetic, include aluminums and **brasses.**

The properties of a metal are determined by the ingredients at the time of manufacture. The amount and kinds of ingredients will determine the composition of a metal and how it will respond in a given application. For example, the addition of the element **silicon** to a batch of **steel** may affect its resistance to corrosion. Following is a list of certain properties all metals have, depending on the added ingredients.

Tensile strength: Resistance of a material to being pulled apart (Figure 3-2)

Impact strength: Resistance of a material to sudden force without fracture

Shear strength: Resistance of a material from two equal forces in the opposite direction

Fatigue strength: Resistance of a material to changing forces or loads

Hardness: Resistance of a material to penetration or denting

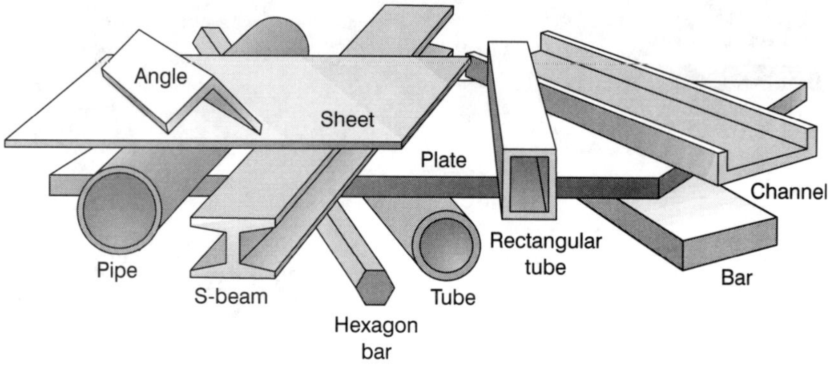

FIGURE 3-1 A variety of structural shapes of metal are required for the many fabricators used in daily living.

FIGURE 3-2 Tensile strength is resistance to being pulled apart.

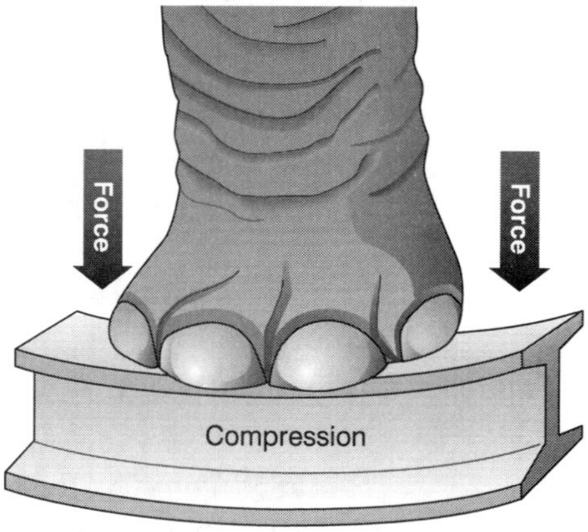

FIGURE 3-3 Compression strength is resistance to being crushed.

Compression strength: Resistance of a material to being crushed (Figure 3-3)

Toughness: Resistance of a material to failure from a constant force

Ductility: Resistance of a material to failure after stretching, twisting, or bending

Steelmaking

Because all metals are affected by welding, some basic understanding of the behavior of steels during welding is important. With the birth of modern steelmaking in the 1850s, people worldwide enjoyed a long relationship with steel. Until recently, developments in techniques of fabrication have concentrated on steel because of its dominance over other metals.

The steelmaking process begins when **iron ore,** limestone, and coke are fed into a blast furnace (Figure 3-4). The type of steel manufactured will depend on the requirements of the customer and, of course, the ingredients that go into the recipe. **Manganese** is an ingredient added to steel to increase its hardness. **Chromium** is added to steel to increase its **corrosion resistance. Molybdenum** is a third element sometimes added as an ingredient to steel to increase its tensile strength.

The four general types of steel are classified as (1) **low-alloy steels,** (2) **high-alloy steels,** (3) tool steels, and (4) **carbon steels.** A low-alloy steel according to the **American Iron and Steel Institute** is a steel in which one of the following amounts is exceeded: 1.65% manganese content, 0.60% silicon content, or 0.60% **copper** content; or, a low-alloy steel is a steel in which a definite minimum quantity of aluminum, **boron,** chromium (up to 3.99%), or a minimum quantity of any other alloying element is added to achieve a desired result.

A high-alloy steel is a steel in which chromium, manganese, or **nickel** content equal 12% or better. Stainless steels are examples of high-alloy steels because their chromium content exceeds 12%. Stainless steels are somewhat more demanding to weld than mild steels because they have (1) a lower melting temperature, (2) a lower heat conductivity, and (3) a higher rate of expansion when heated. As a result, twice as many **tack welds** should be used to reduce **distortion.** A faster **travel speed** is suggested with intermittent welding to promote more rapid cooling.

Carbide precipitation occurs when chromium-nickel stainless steels are held within a temperature range of 800° to 1600° for an extended period. Carbon reacts with chromium in a chemical reaction that does not have corrosion resistance.

Tool steels are used in cutting or forming applications. The carbon steels in this classification contain from 0.60 to 1.5% carbon. The alloy steels in this classification contain in addition to the high carbon, added amounts of chromium, **tungsten,** and molybdenum.

Carbon steels are popular types of steel for welding. The amount of **carbon** in steel, determined during manufacturing at the steelmaking plant, helps to specify its hardness. When the amount of carbon is increased to a batch of steel, the hardness and tensile strength will also increase. However, both the ductility and **weldability** will decrease (Figure 3-5).

Steelmaking begins with iron ore, limestone, and coal taken from the earth.
This diagram shows some major steps in the process.

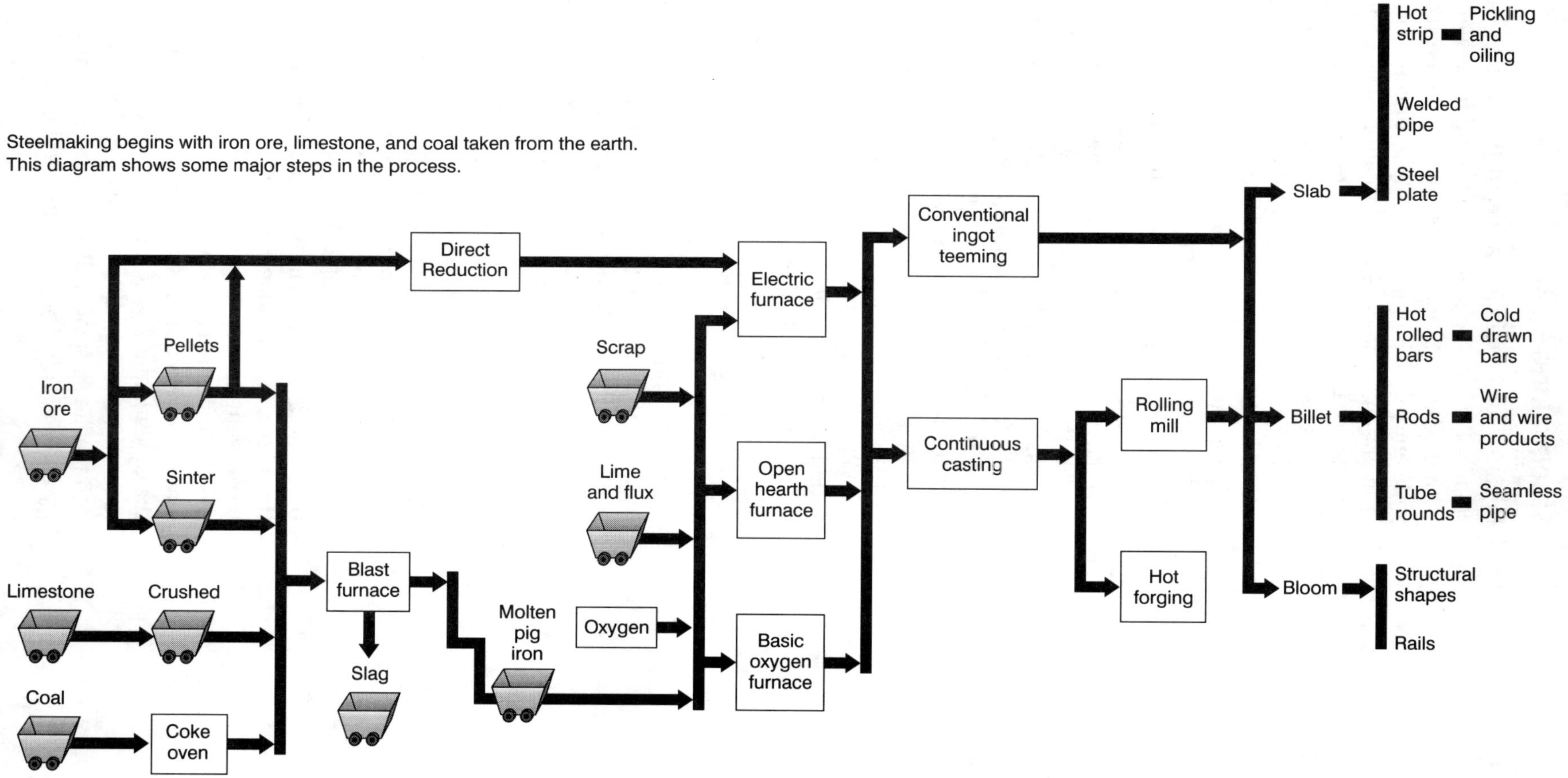

FIGURE 3-4 Steelmaking.

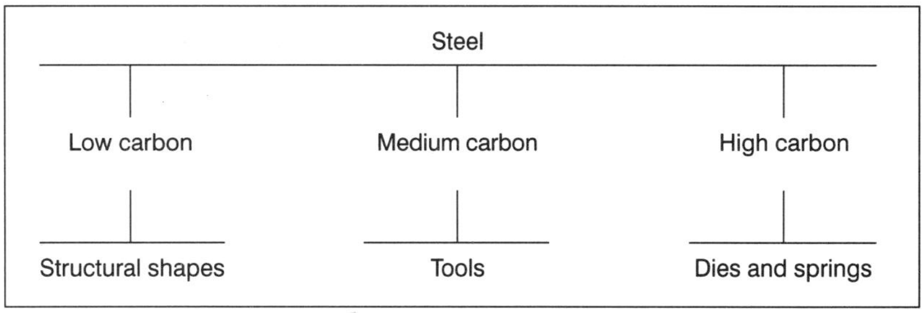

FIGURE 3-5 Uses of different carbon steels.

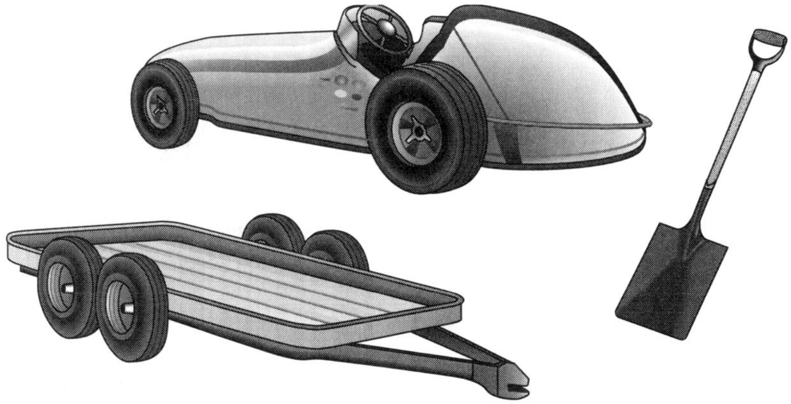

FIGURE 3-6 Varied products made of steel.

Low-Carbon Steel

Low-carbon steel, which is also called **mild steel,** contains 0.1 to 0.3% carbon (which is less than 1%). Low-carbon steel is commonly welded, finding application in construction and fabrication. Low-carbon steel cannot be hardened significantly by heat treatment alone; therefore, low-carbon steel has no application for **tempered** tools such as punches and chisels.

Normally low-carbon steel does not require additional **preheating** or **postheating** procedures to guard against weld failure. Two exceptions that may require preheating relate to the outside ambient temperature and the thickness of the base metal. In short, low-carbon steel is easily welded. Low-carbon steels have application in the structural members used for constructing skyscrapers and bridges. Buildings, piping, pressure vessels, and motor frames are more general examples of products manufactured from low-carbon steel (Figure 3-6). Many products found around the home are constructed of low-carbon steel.

Whereas low-carbon steel is readily welded, such is not the case with other steels. **Medium-carbon steel** (0.3 to 0.45% carbon), **high-carbon steel** (0.45 to 0.65% carbon), **very high-carbon steel** (0.65 to 1.5% carbon), and cast iron (up to 4.5% carbon), all require special care in welding. Table 3-1 shows some applications of different steels. Special care may require preheating and/or postheating procedures. High-alloy steels in which such **elements** as

TABLE 3-1 Applications of different steels

Steel	*Examples of Applications*
Low carbon	structural shapes: channel, angle, bar
Medium carbon	axles, connecting rods, shafts
High carbon	vehicle springs, band saws, anvils
Very high carbon	lathe tools, knives, chisels, punches

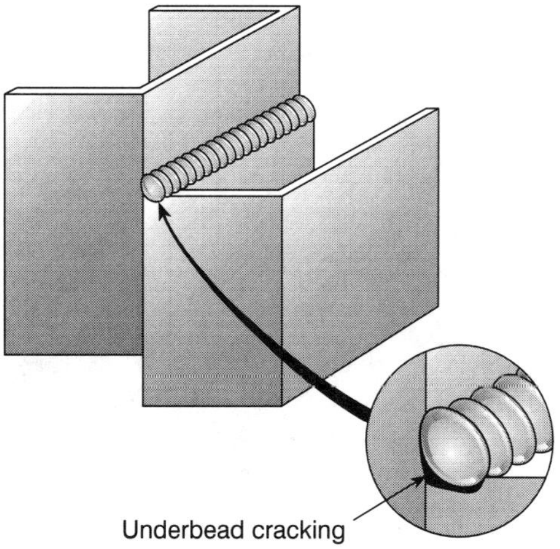

FIGURE 3-7 Underbead cracking results from hydrogen diffused into the heat-affected zone.

chromium are added in a sufficient amount equal to 12% or better may require special care.

The welder must know what particular type of steel is being welded especially when the steel is a type other than low carbon. Appearance is not going to help in recognizing a steel, because most carbon steels look alike. Even if low-carbon steel is the most prevalent, a buyer should learn to depend on a reputable supplier for purchasing structural steel. Buying a known classification of steel under the Unified Numbering System from a supplier will help to prevent **cracking** of a joint and possible weld failure. The welder should know if a particular steel requires special preheat or postheat treatment to prevent problems such as that shown in Figure 3-7.

Identifying Low-Carbon Steel

Low-carbon steel cannot be distinguished from other types of carbon steels by using a magnet. However, a magnet will isolate the carbon steels from other metals that are nonmagnetic such as chromium-nickel stainless steels, coppers, brasses, and aluminums.

Low-carbon steel can sometimes be identified from other types of **carbon steel** by filing or using a chisel. A file or a chisel made for use on steel will bite easily into low-carbon and medium-carbon steel to remove metal chips (see Table 3-2).

Apply pressure with the file to unknown steel on the forward stroke only (Figure 3-8). The file and the chisel will remove chips with difficulty or not at all as the percentage of carbon content increases into high-carbon and very high-carbon steel.

ARC FLASH

All metals have different melting points.

Steel 0.4–0.7% carbon 2500°

Stainless steel (nickel) 2550°

Steel 0.15–0.4% carbon 2600°

Stainless steel (low carbon) 2640°

Steel (less than 0.15% carbon) 2700°

TABLE 3-2 Recognizing metals by filing or chiseling

Base Metal	*Use of Chisel*	*Use of File*
copper	chips easily	bites easily
brass	chips easily	bites easily
aluminum	chips easily	bites easily
low-carbon steel	continuous chip	bites easily
medium-carbon steel	chips easily	bites with pressure
high-carbon steel	difficult to chip	does not bite

FIGURE 3-8 Use of a file to aid in determination of carbon content.

REVIEW

1. Name the type of steel that contains the following:
   a. 1% manganese, 0.40% silicon content, or 0.55% copper
   b. 3% chromium
   c. 13% chromium
   d. 1% carbon
2. How does the addition of carbon affect steel?

Lab Exercises

1. Take pieces of steel supplied by your instructor and, using a file or a chisel, attempt to identify the types of steel?
2. Identify some of the structural shapes in which metal is made that are found in your shop.

STEEL PLATE

Preparation for Welding

A structural quality low-carbon steel will be needed for the exercises in this book. According to standards established by the **American Society for Testing and Materials (ASTM),** A36, A283, A570 are three steels suitable for welded structures. The recommendation is for a general grade structural steel that can be readily welded without added preparation beyond those presented in this unit.

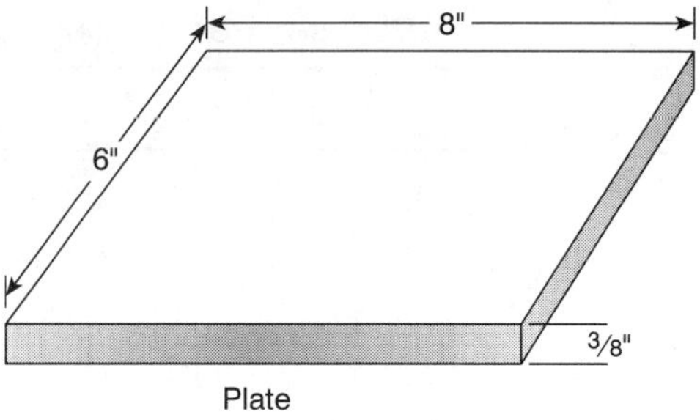

FIGURE 3-9 Plate dimensions for many of the exercises in this book.

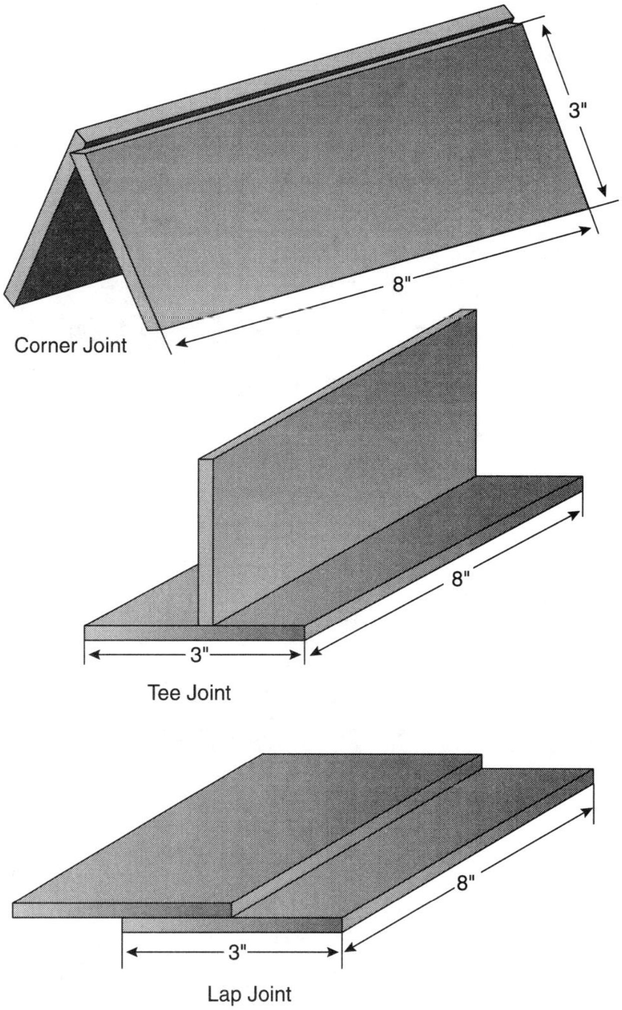

FIGURE 3-10 Corner joint, tee joint, and lap joint dimensions for welding.

Plate is one structural metal shape. Plate is defined as over 3⁄16 inch thick and over 6 inches wide. Sheet is another structural metal shape. Sheet is 3⁄16 inch thick and less, but over 6 inches wide. A thickness of 16-gauge sheet steel to 3⁄8-inch-thick steel plate is recommended for all exercises. The thinner the material the more distortion that may be experienced. The padding exercise requires plate 3⁄8 inch thick, 6 inches wide by 8 inches in length (Figure 3-9).

FIGURE 3-11 Torch designed for oxyacetylene cutting.

Strip is a structural shape $\frac{3}{16}$ inch thick and under, and up to 6 inches wide. **Flat bar** is another structural shape metal. Flat bar is over $\frac{3}{16}$ inch thick and up to 6 inches wide, commonly available in 20-foot lengths. Several lengths 8 inches long should be cut out of 3-inch strip and flat bar for the **corner joint, lap joint,** and **tee joint** exercises (Figure 3-10). Angle iron, 3 inches, can be used as a substitute for flat bar in the corner joint exercise.

The first step is to cut steel to the dimensions necessary for the exercises. Strip and flat bar are easier to handle because it comes in 6-inch widths that can be cut 8 inches long. Unlike plate, strip or flat bar requires a single cut to prepare for the welding exercises presented in this book. Some metal suppliers will cut material to size as requested. With training, you can learn to use a band saw, **oxyacetylene** torch, air carbon arc, or plasma cutting torch to do the job (Figures 3-11 and 3-12).

Mill scale and any rust should be removed from the surface of the joint where the welding will be completed. A handheld grinder is a useful tool for this purpose and belongs in any shop where welding is done. A versatile tool, the grinder can be used for quickly grinding away welds. Grinding can add to the appearance of any finished fabrication project (Figure 3-13).

FIGURE 3-12 The horizontal band saw is an important tool for metal cutting. It can be used to cut coupons for weld testing.

FIGURE 3-13 The grinder is an important shop tool, used to clean metal surfaces for grinding or beveling edges.

REVIEW

1. How is plate different from flat bar?
2. What are some tools used for cutting metal to size?

Lab Exercises

1. Research and provide the ingredients for one of the following steels: A36, A283, or A570.
2. Become familiar with safe operation and maintenance of a particular tool that will be used in your welding. Hint: Read the operator's manual for one of the following tools: grinder, band saw, or air carbon arc torch.
3. Based on information provided by the instructor, write the safety precautions to be followed for the tool(s) used in cutting metal plates to size for welding.

Composition and Identification

Gas metal and flux cored arc welding require electrodes. The electrode is the filler metal added to the joint (Figure 3-14). Electrodes must be matched to the base metal being welded. The American Welding Society (AWS) has developed filler metal specifications. These specifications are used by manufacturers in setting the chemical composition, physical properties, and classification of their electrodes. These specifications are also used in establishing the **welding procedures** for testing to determine the application of a given electrode. The properties of the weld are affected by many factors including electrode choice.

The electrode is designed to be fed continuously through the gun at a high rate of speed. The spool of wire that can weigh from 1 to 60 pounds turns freely to uncoil the wire smoothly without interruption. Larger drums of wire that can weigh up to 1,000 pounds are also available.

A tubular wire is the electrode design for FCAW. A hollow in the continual wire is filled with a granular **flux,** which is sealed by the pressure of compression to its diameter (Figure 3-15). The flux melts during the welding process and produces a gaseous shield that protects the weld from the air. Ingredients in the flux attach

FIGURE 3-14 Electrode filler metal.

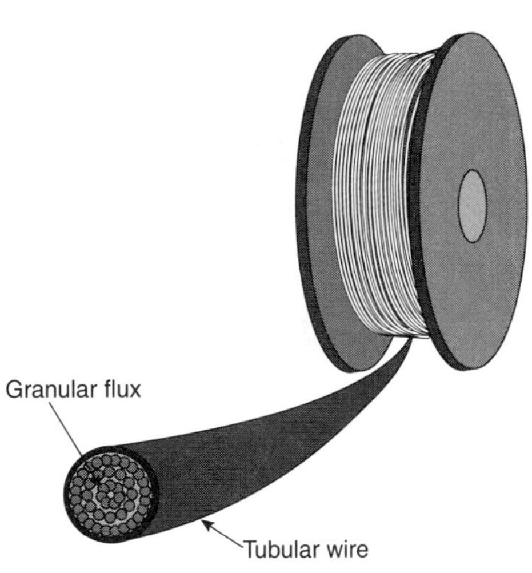

FIGURE 3-15 A tubular wire is filled with a granular flux.

themselves to impurities in the molten weld pool. The flux also determines penetration into the base metal and helps to stabilize the arc to make welding easier. The lighter flux rises to the surface of the weld pool created by the arc, forming slag. Slag which protects the weld bead as it cools must be chipped off with a slag hammer before laying another weld bead.

Electrodes are identified by the numbering classification system developed by the AWS. This numbering system makes it possible for the welder to identify the wire electrode. The numbering system is different, depending on the base metal and whether the welding is by GMAW or FCAW. For example, a base metal of carbon steel, alloy steel, or perhaps aluminum fits under a different specification.

The following AWS specification A5.18 is for carbon steel filler metals used by gas metal arc welding (GMAW).

E70S-2

The *E* indicates an arc welding electrode. The first two digits *70* (multiplied by 1000 = 70,000) refer to the minimum tensile strength of the steel filler metal per square inch. The *S* identifies the wire as a solid wire. A number *2* classifies the wire by its chemical composition and the weld's physical properties. In this case the number *2* indicates that a wire electrode can be used to make quality welds on rusty surfaces with argon/oxygen, argon/carbon dioxide, or carbon dioxide.

The following AWS specification A5.20 is for carbon steel filler metals used by flux cored arc welding (FCAW).

E70T-2

The *E* indicates an arc welding electrode. The first digit *7* (multiplied by 10,000 = 70,000) refers to the minimum tensile strength of the steel filler metal per square inch. The second digit *0* refers to the **welding positions.** Zero indicates welding in the **flat position** and the **horizontal position.** The *T* identifies the wire as a tubular wire. A number *2* classifies the wire by its chemical composition, shielding gas, current type, and the weld's physical properties. In this case a number *2* indicates a wire electrode that can be used to make quality welds in a single pass on rusty surfaces, gas shielding by carbon dioxide, and direct current electrode positive (DCEP). Unit 8 will cover electrode wires in greater detail.

SAFETY REMINDER

Turn off the switch when leaving the power source unattended. Avoid the possibility of the trigger on the gun energizing the wire, which would cause flowing wire to arc and create a situation for the possibility of an explosion or a fire.

REVIEW

1. How is a tubular wire different from a solid wire?
2. How is the flux related to a shielding gas?

Lab Exercises

1. List the AWS numbering of wires used in your shop.
2. List the shielding gases used in your shop.

REVIEW QUESTIONS FOR UNIT 3

Multiple Choice

Choose the *best* answer to complete the statement.

1. Nonferrous metals:
   a. Are magnetic.
   b. Include carbon steels.
   c. Include low-alloy steels.
   d. Include aluminums.
   e. Are none of the above.
2. Impact strength is resistance to:
   a. Failure from a constant force.
   b. Penetration.
   c. Sudden force.
   d. Changing forces.
   e. None of the above.
3. Compression strength is resistance to:
   a. Being pulled apart.
   b. Failure from a constant force.
   c. Deformation after stretching, twisting, or bending.
   d. Denting.
   e. None of the above.
4. Ductility is resistance to:
   a. Being crushed.
   b. Equal force in opposite direction.
   c. Changing loads.
   d. Failure after stretching, twisting, or bending.
   e. None of the above.
5. Chromium is:
   a. Magnetic.
   b. Corrosion resistant.
   c. An ingredient of carbon steels.
   d. Answers a. and b. above.
   e. Answers b. and c. above.
   f. All of the above.

6. ____________ are high-alloy steels.
   a. Aluminums
   b. Brasses
   c. Stainless steels
   d. Carbon steels
   e. None of the above
7. As the carbon in a batch of steel increases, (what happens next?)
   a. Hardness decreases.
   b. Weldability increases.
   c. Hardness increases.
   d. Ductility decreases.
   e. Answers b. and c. above.
   f. Answers c. and d. above.
8. Stainless steels when compared with carbon steels have:
   a. Lower melting point.
   b. Lower heat conductivity.
   c. A higher rate of expansion.
   d. All of the above.
   e. None of the above.
9. Carbon content of tool steel is:
   a. 0.3% to 0.45%.
   b. 0.1% to 0.3%.
   c. Up to 4.5%.
   d. 0.45% to 0.65%.
   e. 0.6% to 1.5%.

Short Answer

1. Explain what determines the composition of metal in its manufacturing process.
2. List the three furnaces that are fed from the blast furnace.
3. List the structural shapes made from billets.
4. List the structural shapes made from slabs.
5. Describe the American Iron and Steel Institute.
6. Explain when steel becomes a high-alloy steel.
7. Explain why carbide precipitation is bad.
8. Tell which has more carbon, cast iron or very high-carbon steel. Explain.
9. Define distortion.
10. Explain the basis of the different numbering systems for electrode wires.

ADDITIONAL ACTIVITIES

Crossword Puzzle

ACROSS
1. Organization dedicated to welding (abbreviation).
2. Structural metal shape.
3. Welding position.
4. Fuel fed to the blast furnace.
5. Product of the flux.
6. Oxides on the surface of metal.
7. Helps to specify a steel's hardness.
8. Measurement of defiance to being pulled apart.

DOWN
1. Welding that meets a credible standard of excellence.
2. High-alloy steel.
3. Filler metal.
4. Hollow wire.
5. Shielding gas.
6. Can be used for metal identification.
7. Added to steel to increase its hardness.
8. Steel suitable for structural welding.

Writing Skills

1. Write a paragraph using the following words: *stainless steel, high-alloy, mild steel, low carbon, chromium, carbon,* and *rust.*
2. Explain the differences in the numbering systems for the wires used in gas metal arc welding and flux cored arc welding.

Math

$>$ (greater than)	or	$<$ (less than) ???
1 ? 0.1		0.1% ? 0.1
1 $>$ (is greater than) 0.1		0.1% (0.001) $<$ (less than) 0.1

Solve for greater than or less than.

1. 1% ___ 0.1 **2.** 0.3% ___ 1% **3.** 0.2 ___ 0.45%

4. 0.3% ___ 0.45% **5.** 0.65%___ 0.65 **6.** 0.65% ___ 0.45%

7. 0.01 ___ 1.0% **8.** 0.65% ___ 1.5% **9.** 4.5% ___ 0.04

10. 1% ___ 0.045

A person is considered intoxicated in many states when the blood-alcohol concentration reaches 0.1%; above 0.5% concentrations may cause death. What is the status of individuals with the following blood-alcohol concentrations?

11. 0.01% **12.** 0.085% **13.** 0.1%

14. 0.35% **15.** 0.75%

Puzzlers

1. Explain why the welding of thin metal is more subject to distortion.
2. Explain how to avoid carbide precipitation in the welding of stainless steel.
3. Which is of greater hardness, steel or glass? Explain.

UNIT 4

Fundamentals of Welding

ACT OF WELDING

Beginning the Arc

Every weld begins by squeezing the trigger to energize the electrode. It may be necessary to perform some trial-and-error welding on scrap metal to set the parameters for welding. Be careful to begin the arc in the welding area. Stray arc strikes outside of the weld produce brittle areas which may crack under **stress.** A careless arc strike can damage parts originally not the focus of the welding. The instructor will most certainly reject any **weldment** with several arc strikes outside the welding area (Figure 4-1).

Even before squeezing the trigger, the welder makes many decisions according to the welding variables. These variables include the size and type of electrode, shielding gas, the type and thickness of the base metal, the type of joint, the position for welding, and the metal transfer (Figure 4-2). Once the variables are selected, adjustments are made at the power source to set the voltage and wire speed. Beyond these parameters the welder must be concerned about such welding techniques as **electrode extension, travel angle, work angle,** and travel speed.

Travel Angle

The two angles important to the welder in the act of welding are the travel angle and the work angle. Travel angle is the angle of the gun in relation to the direction of travel along the joint. The gun should be positioned for welding in the flat and horizontal positions with a travel angle from 15° to 45° (Figure 4-3). The travel angle will be discussed throughout the welding exercises of this book. Decreasing the travel angle will increase the root penetration into the base metal. Increasing the travel angle will not only decrease root penetration, but will also flatten the weld bead.

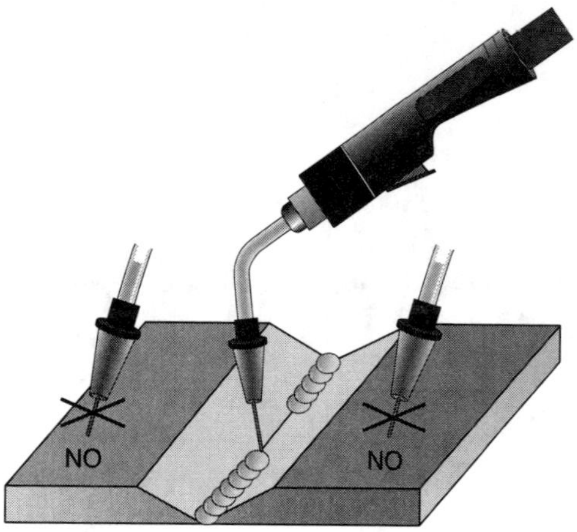

FIGURE 4-1 Keep arc strikes in the weld.

FIGURE 4-2 Welding involves knowledge.

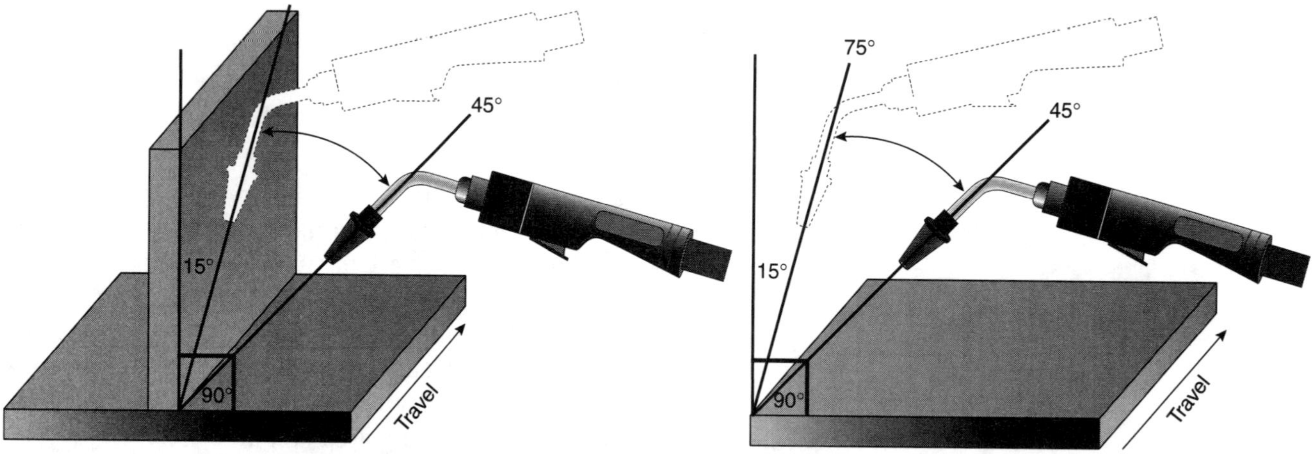

FIGURE 4-3 Travel angle of the gun is important.

Work Angle

The work angle, the second of the two important angles, is the angle of the electrode in relation to the base metal. The work angle will vary according to the specific job. Welding on plate in the flat position, the 180° angle is bisected with a 90° work angle (Figure 4-4). Welding a tee joint in the horizontal position, the 90° angle is bisected with the work angle adjusted according to the sequence of the weld bead (Figure 4-4). Maintaining the correct work angle will help to avoid undercut along the weld (Figure 4-5).

Direction of Travel

There are two methods of travel when operating the gun. With the **backhand** welding technique, the nozzle of the gun is pointed back at the weld pool opposite the direction of travel (Figure 4-6). Backhand welding causes more root penetration into the base metal because the heat of the arc is directed back on the weld bead. Though backhand welding provides less visibility because

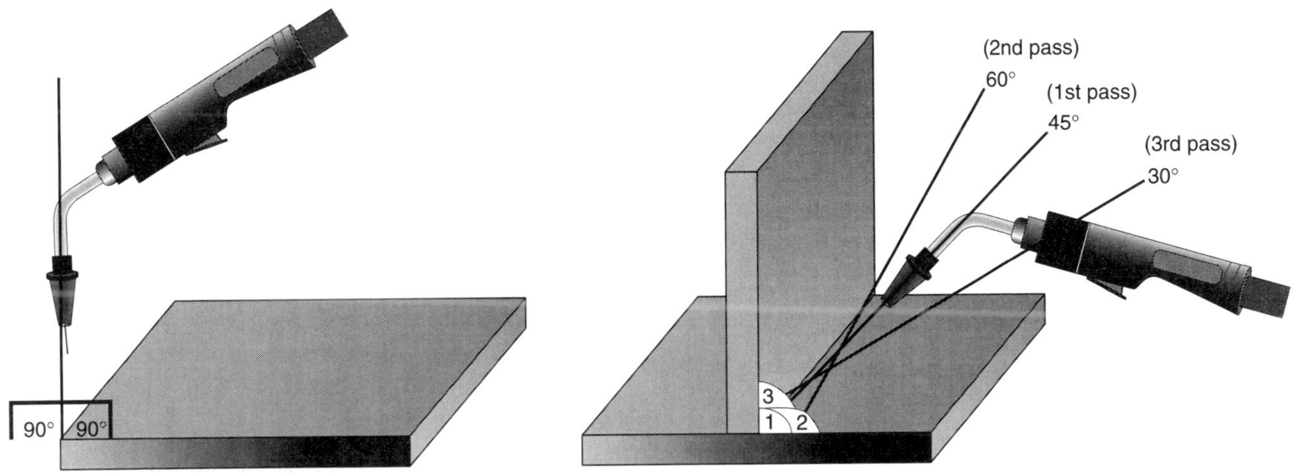

FIGURE 4-4 Work angle is important to avoid undercut.

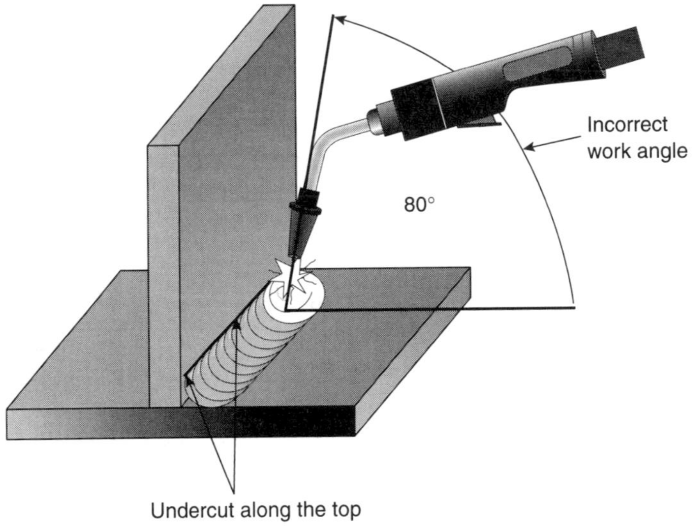

FIGURE 4-5 Guard against undercut.

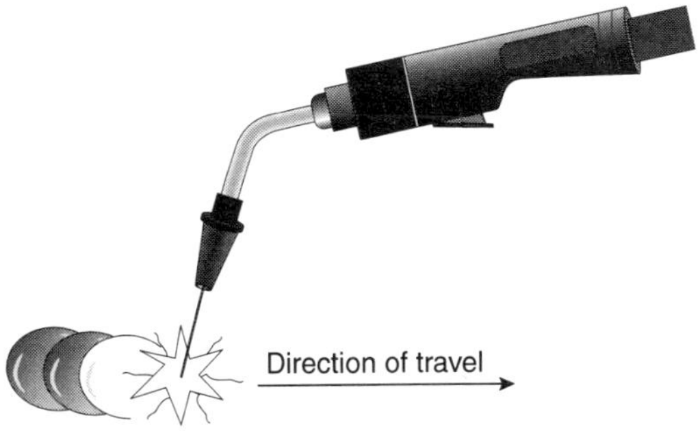

FIGURE 4-6 Backhand welding technique.

the gun is positioned over the joint to be welded, during GMAW the shielding gas offers better protection of the weld in windy or drafty conditions.

With the **forehand** welding, the nozzle of the gun is pointed at the leading edge of the weld pool in the direction of travel (Figure 4-7). Forehand welding causes less root penetration into the base metal because heat of the arc is used to raise the temperature of the base metal along the joint. Forehand welding offers good visibility of the joint to be welded, but the shielding gas is more likely to be affected by windy or drafty conditions because the gas flow is not concentrated on the cooling weld bead.

> **ARC FLASH**
>
> Replace worn insulators to prevent arcing off the nozzle of the gun, which can damage the nozzle and affect the quality of the work.

Electrode Extension

Because the arc length is controlled by the adjustments of the voltage and wire speed, the welder must be concerned with the *electrode extension*. The electrode extension is the distance between

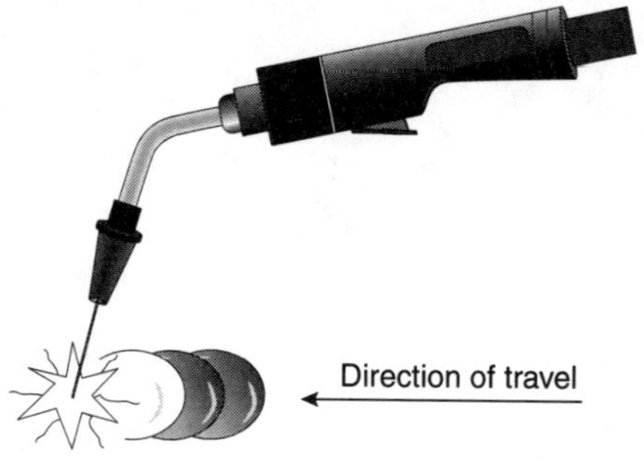

FIGURE 4-7 Forehand welding technique.

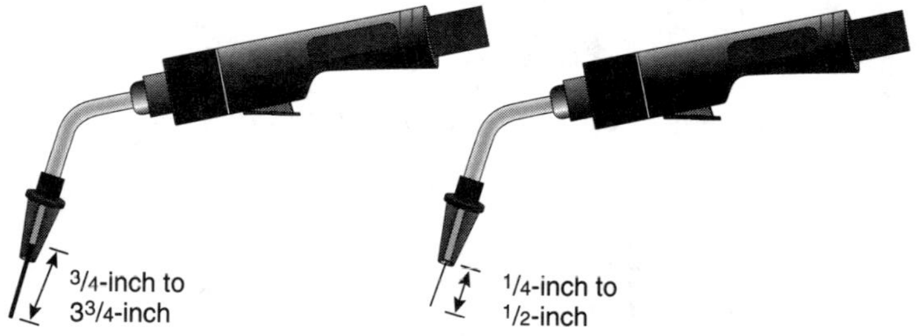

FIGURE 4-8 The electrode extension for GMAW is from ¼ to ½ inch. The electrode extension for FCAW is from ¾ to 3¾ inches.

the end of the contact tube and the tip of the unmelted wire (Figure 4-8). Changing the position of the gun for a longer electrode extension creates a greater electrical resistance. This causes a slight voltage drop from the end of the wire to the base metal to reduce the rate of melting. The power source then responds by lowering the welding current. The weld bead becomes narrower with a higher weld face. Changing the position of the gun to a shorter electrode extension creates less electrical resistance. This will increase the welding current to accelerate the rate of melting. The weld bead will then spread out.

The recommended length of the electrode extension for GMAW is from ¼ inch to 1 inch; and for short circuiting transfer, ¼ inch to ½ inch. The recommended length of electrode extension for FCAW is from ¾ inch to 3¾ inches. Remember that too long an extension causes an unstable arc with excessive spatter. Too short an extension will arc the electrode back into the nozzle, possibly fusing the wire onto the contact tube.

Travel Speed

Travel speed is the rate of movement for the weld pool along the joint (Figure 4-9). When travel speed slows, the amount of deposited filler metal per linear inch increases. When travel speed quickens, the amount of deposited filler metal per linear inch decreases. Travel speed should be coordinated with the wire speed adjustment. The proper travel speed must be maintained to control the amount of root penetration into the joint. Moving too fast,

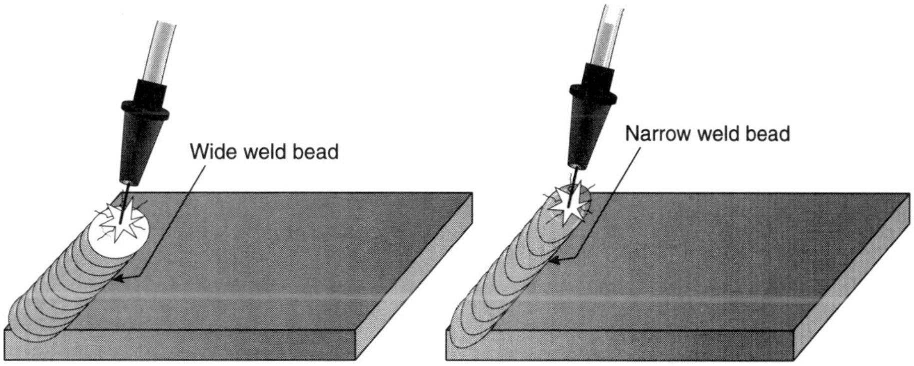

FIGURE 4-9 Travel speed determines the width of the weld bead and root penetration into the joint.

the root penetration decreases. Moving too slow, the root penetration can result in melt-thru on thin metal, or the weldpool will flow into the joint ahead of arc on thicker metal.

Lab Exercises

1. Make a list of the decisions that must be made before fabricating a gas tank.
2. Make a list of the decisions that must be made before repairing a trailer.

METAL TRANSFERS

Short Circuiting Transfer

With the short circuiting transfer type of GMAW, the wire electrode as it makes contact with the weld pool causes a rise in amperage. The wire heats up and small metal droplets are pinched off the end, igniting the arc. The arc is then once again extinguished by the weld pool to begin the cycle again (Figure 4-10). This cycle repeats itself up to 200 times per second. Short circuiting transfer has the lowest amperage settings of the metal transfers, making it advantageous on thin gauge metals and to bridge large openings. Much spatter occurs with this process, which is caused by each separation of the metal droplets off the end of the wire. The fast-freeze weld pool makes short circuiting transfer suitable for all welding positions.

Globular Transfer

With the globular transfer type of GMAW, a transfer of drops (larger than the wire's diameter) occurs across the arc (Figure 4-11). As each molten drop grows in size, it drops away by short-circuiting on the base metal, or by gravity. When the drops separate by short-circuiting, the spatter can be great. Using amperage settings slightly higher than short circuiting transfer, globular

FIGURE 4-10 The cycle of short circuiting transfer repeats itself 200 times per second.

FIGURE 4-11 With globular transfer, each drop grows in size as it drops away.

transfer is limited in applications to the flat position where the molten drop that forms on the tip drops away under the pull of gravity.

Spray-Arc Transfer

With the spray-arc transfer type of GMAW, a transfer of tiny droplets (hundreds per second) occurs across the arc (Figure 4-12). This process requires higher voltage settings than either short circuiting transfer or globular transfer, and carbon dioxide shielding gas cannot be used. A mixture of 98% argon to 2% oxygen is a commonly used shielding gas for welding mild steel and stainless steel; 100% argon is used for welding aluminum. Recent development in technology of pulsed GMAW has overcome limitations that previously restricted spray-arc transfer to the flat and horizontal positions. The newer power sources allow welding to be controlled and completed on base metals of various thicknesses in all welding positions.

Two Methods of Flux Cored Arc Welding

Flux cored arc welding applies the characteristics of having a flux in the self-shielded method (FCAW-S). Flux cored arc welding also combines a flux with a shielding gas in the gas-shielded method (FCAW-G). Both methods protect the weld bead as it forms. The self-shielded method uses a tubular wire with the flux

FIGURE 4-12 Spray-arc transfer produces hundreds of small droplets that cross the arc.

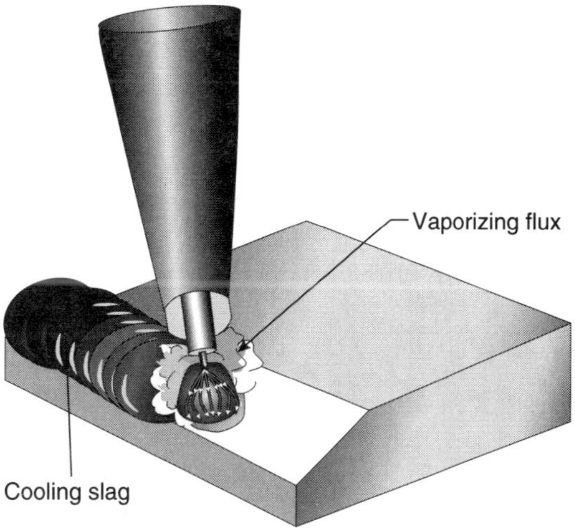

FIGURE 4-13 Self-shielded FCAW.

contained inside (Figure 4-13). The vaporizing flux is designed to deal with oxygen and nitrogen from the air that saturate the weld pool. The slag covering that forms on top of the molten pool of liquid metal protects the weld bead as it cools. The gas-shielded method uses a gas, possibly carbon dioxide or a mixture of carbon dioxide and argon, in addition to a flux-contained tubular wire.

Self-shielded method is sometimes used in applications when shielded metal arc welding could have been used, but is not economical. Self-shielded flux cored arc welding (FCAW-S) wires require simple guns and are suited for outside welding; they handle the weather better than wires designed for shielding gases, although they produce more fumes. Gas-shielded flux cored arc welding (FCAW-G) wires are used in applications when joint design requires a narrow weld bead with deep root penetration characteristics.

Suggestions to Improve Welding Technique

Following are tips to remember to improve welding skills.

1. Always make sure good electric contact is at the workpiece connection as close to the welding as possible.
2. Relax and find a comfortable position. Do not grip the gun too tightly. Quality welding depends on staying loose, to reduce fatigue.
3. Stay aware of both the work angle and the travel angle during the act of welding.
4. When using both hands, support one hand with the other by gripping the forearm.
5. Position the gun so as to keep the weight of the power cable from causing undo stress.
6. When gas metal arc welding, protect the welding from strong drafts or wind to keep from blowing the shielding gas away.

7. When flux cored arc welding, take extra care to eliminate fumes and remove the slag as required before completing another pass.
8. For safety, always chip slag away from the body and keep the helmet down to prevent slag from flying into your face.

REVIEW

Connect each term with its description.

1. parameters	a. work angle, travel angle, etc.
2. short circuiting arc transfer	b. narrow weld bead with high weld face
3. technique	c. increases root penetration
4. travel angle	d. size, type of electrode, etc.
5. slow travel speed	e. thin gauge thicknesses, bridge large openings
6. decrease travel angle	f. electrode position to base metal
7. work angle	g. decreases root penetration
8. method of travel	h. increase deposition of filler metal
9. forehand welding	i. backhand welding
10. increase electrode extension	j. direction of travel

ADDITIONAL ACTIVITIES

Word Find

Directions: Circle the technical words listed below.

E	L	X	N	A	L	P	D	Q	E	A	J	S	METAL TRANSFER
L	R	R	U	L	S	R	T	U	O	O	E	H	ARGON
G	X	A	N	L	N	T	I	A	I	O	L	I	FLUX
N	P	G	U	N	F	V	O	L	T	A	G	E	VOLTAGE
A	W	L	N	P	A	A	G	I	R	X	F	L	WORK ANGLE
L	O	E	D	A	R	K	R	T	S	E	O	D	TRAVEL ANGLE
E	R	C	E	Z	G	A	L	Y	E	S	R	I	SPATTER
V	K	T	R	D	O	E	G	Q	P	P	E	N	SHIELDING GAS
A	A	N	C	G	N	A	N	U	A	M	H	G	QUALITY
R	N	I	U	M	L	R	I	A	Y	M	A	G	WIRE SPEED
T	G	O	T	S	P	A	T	T	E	R	N	A	UNDERCUT
M	L	J	W	I	R	E	S	P	E	E	D	S	JOINT
R	E	F	S	N	A	R	T	L	A	T	E	M	SLAG
													GUN
													FOREHAND

Writing Skills

1. In a paragraph, describe the difference between the short circuiting transfer and the spray-arc transfer.
2. In a paragraph, describe the difference between the self-shielding method and the gas-shielding method.

Math

Degrees and Angles

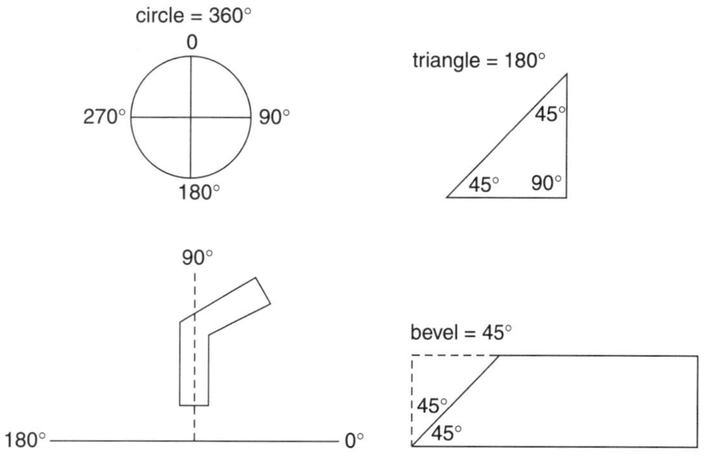

Give the approximate degrees represented in the shaded area for each circle.

1. 2. 3. 4. 5. 6. 7.

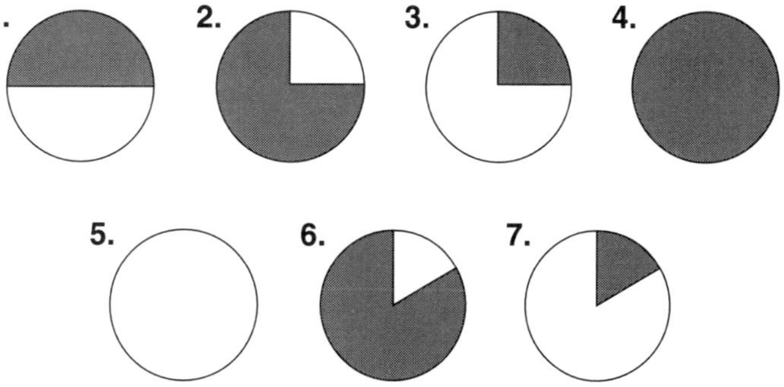

What is the amount of material to be removed (answers in degree) from the illustrations in questions 8 through 11?
What is the degree of the bevel in questions 12 and 13?

8. 9. 10. 11. 12. 13.

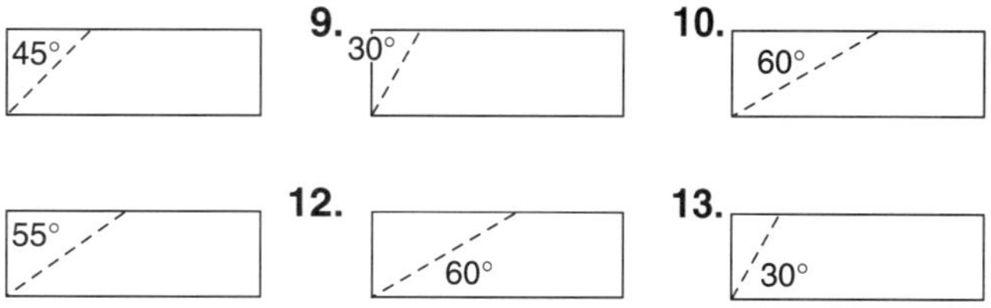

What is the angle for the gun in the following illustrations?

14.

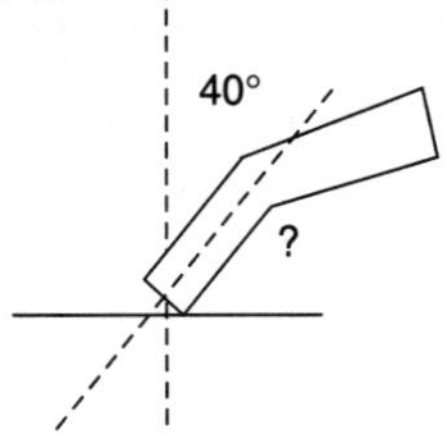

15.

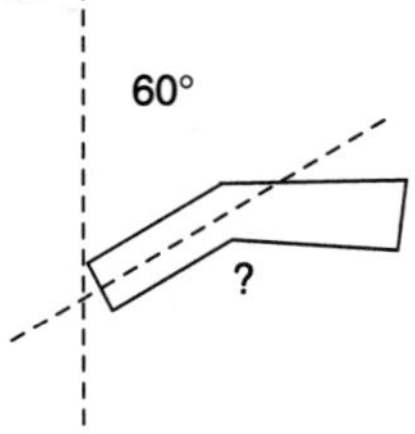

16.

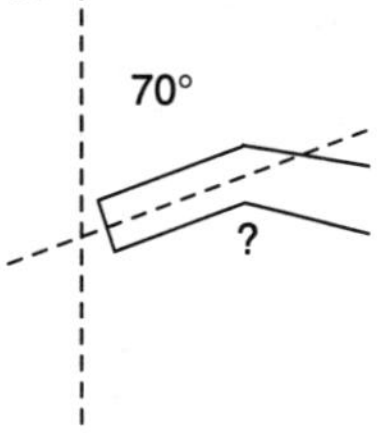

17.

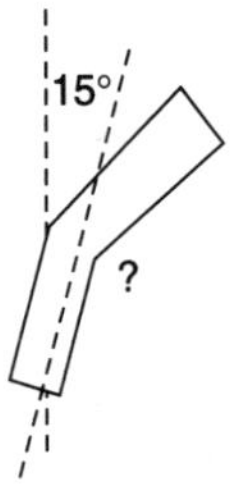

18.

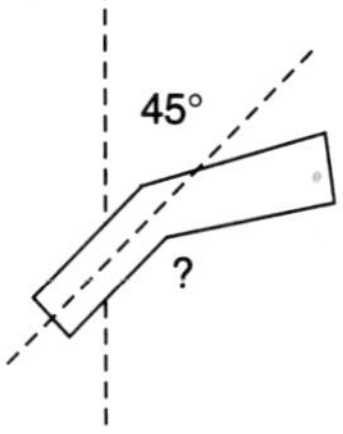

Puzzlers

1. What metal transfer process would you use to attach a partial replacement fender to your automobile, and why?
2. The job calls for fabricating a bucket for a front-end loader. Which welding process would you use and why?

UNIT 5

Common Welding Problems

THE WELD BEAD

Two Basic Methods

The weld bead is deposited metal from a single pass (Figure 5-1). Both GMAW and FCAW create a higher concentration of energy for deeper root penetration at the same amperage setting as shielded metal arc welding.

The **stringer** and the **weave** are the two basic methods for depositing filler metal. The stringer is a weld bead with little to no sideward movement of the gun, usually within a width of 1¼ times the thickness of the base metal (Figure 5-2). By using the stringer the welder can maintain a uniform weld bead in a straight line to focus the heat for maximum root penetration. Laying stringers should be the method of choice for the basic exercises in this book. Stringers allow the welder to concentrate on making weld beads in a straight line. Not all joints are welded using stringers if it becomes necessary to bridge a gap or do **buildup** using a sideward motion. Stringers are not always the choice when performing **uphill** welding in the **vertical position.**

The weave is a weld bead made from oscillation of the electrode from side to side (Figure 5-3). One problem in making a weave is the

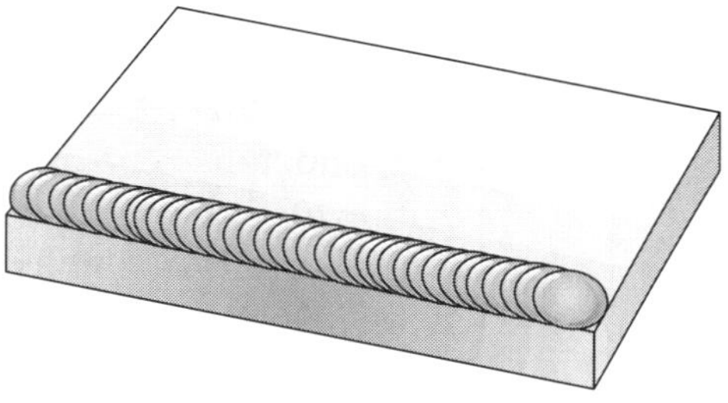

FIGURE 5-1 The weld bead positioned along the edge of the plate.

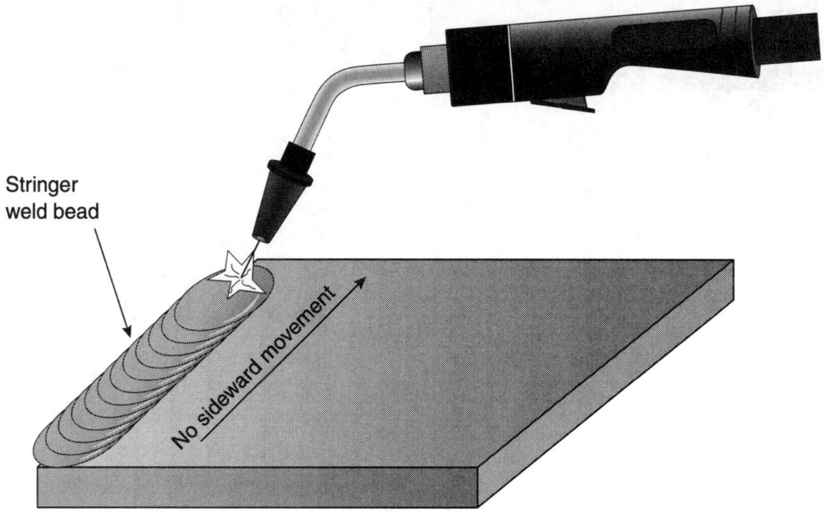

FIGURE 5-2 The stringer weld bead within a width of 1¼ times the thickness of the base metal.

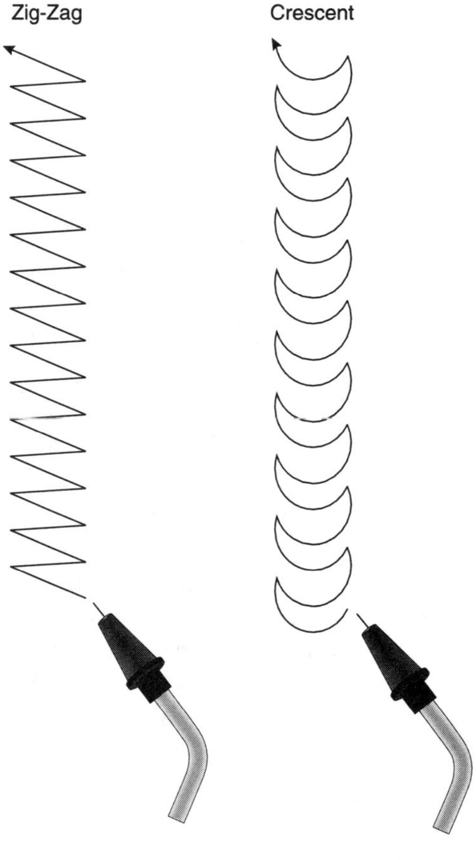

FIGURE 5-3 Two examples of weave weld beads.

welder's failure to pause on the sides of the weld bead long enough to avoid **undercut.** When the welder fails to pause, the weld bead has a tendency to be heavier in the middle. Welding in the vertical position tends to produce a sag or "grapes" in the middle of the weld bead (Figure 5-4). Until you learn to read the weld pool, one remedy to ensure the necessary pause is to count "one thousand one" on each side. This pause provides enough time to avoid undercut

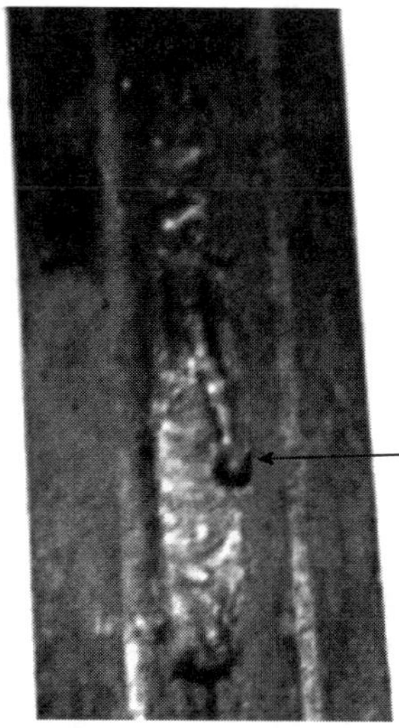

FIGURE 5-4 Grapes are a common problem in the vertical position when the welder loses control of the weld pool.

and make a smooth weld bead with even fill along the entire length of the weld. The weave tends to disperse the heat that can restrict root penetration into the base metal, which is more of a problem with shielded metal arc welding. Both GMAW and FCAW produce greater current density in the arc for deeper root penetration.

Welding demands control by keeping the hands steady. With practice, you will develop some consistency in welding so that all the weld beads will begin to look similar, which occurs once you achieve consistent travel speed along the joint with every weld bead.

Failure of the Weldment

Unfortunately, the outcome from welding is not always positive. Every welder should become familiar with certain problems and learn how to make the necessary corrections (Figure 5-5).

Several problems can cause failure of the weldment. Incomplete root penetration can result from too fast a travel speed, too low an amperage and wire speed setting, or the improper preparation of the assembly for welding. In the act of welding, the welder may be unaware that the root penetration is incomplete. The appearance of the weld indicates that the travel speed is too fast or that the amperage and wire speed setting are too low.

If the travel speed is too fast, the weld bead has a high **weld face.** The weld face is the surface of the weld. With too fast a travel speed, the weld bead is narrow with the weld pool ripple patterns coming to points (C). If the amperage or wire speed setting is too low and the voltage is too high, the width of the weld bead is wide with a high weld face (A), which can cause molten metal to be thrown away from the weld pool as spatter. Spatter refers to the balls of metal forming on and around the weld bead. There is also a tendency for the wire to burn back onto the contact tube. Improper preparation of the assembly for welding can include a failure to clean the base metal properly, or poor joint design (Figure 5-5).

Undercut is a groove left unfilled by deposited metal along the **weld toes** of the weld bead (Figure 5-6). The weld toes run the

FIGURE 5-5 (A) Wire speed low, voltage high; (B) Wire speed too high; (C) Travel speed too fast; (D) Acceptable weld bead; (E) Travel speed too slow; (F) Porosity.

length of the weld bead to establish the boundaries of the weld face. Undercut can be the result from a voltage or wire speed setting that is too high (B), too fast a travel speed (C), too long an electrode extension, or improper gun angle. Undercut will occur if the welder does not pause at the end of the weld bead to fill the **crater.** A crater is an undesirable depression left at the end of a weld bead (Figure 5-7). Either insufficient post flow of gas or removing the gun and its gas shielding before the crater is solid will cause crater defects.

If the travel speed is too slow (E), the welding arc is prevented from melting the base metal adequately. The weld pool will flow onto the unmelted base metal resulting in **overlap** as the weld pool becomes too large. It is important to increase the travel speed by keeping the arc at the leading edge of the weld pool, or reduce the wire speed.

Porosity refers to pinholes (trapped gas) that form in the weld bead (F). Porosity can result if shielding gas flow is too low or too high. Windy or drafty conditions and a clogged nozzle are other causes. Porosity can also result from **arc blow.**

Incomplete fusion will occur if the travel speed is too fast (C), or when the travel speed is too slow and the weld pool is al-

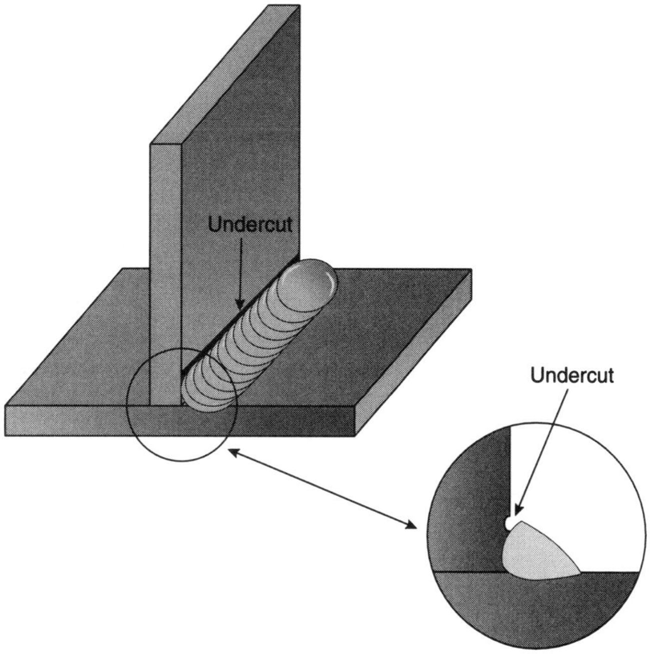

FIGURE 5-6 Undercut has several causes.

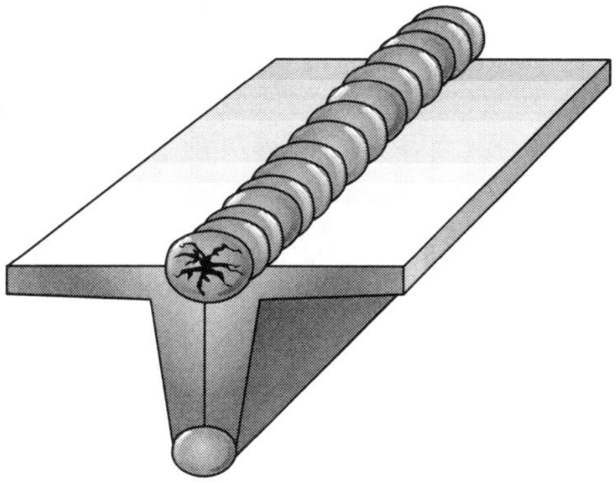

FIGURE 5-7 Crater cracks are another common welding problem.

ARC FLASH

Use antispatter products sparingly. These products can create a greasy buildup on the inside of the liner, combining with impurities to cause a restriction in the movement of the wire.

lowed to flow ahead of the welding arc (E). Incomplete fusion can also result when the work angle of the gun is such that the weld bead ends up primarily on one member of the joint (Figure 5-8).

Melt-thru is incomplete fusion, causing excessive root penetration. It is the result of too high an amperage or wire speed setting and too slow a travel speed (Figure 5-9).

Slag

Slag is nonmetallic material that can affect weld quality; it is the remaining residue when the tubular wire electrode of FCAW melts into the weld pool. When slag forms on the weld bead it can usually be removed easily upon cooling. The slag helps to protect the weld pool as it cools to keep contamination from the air being drawn into the weld as it becomes solid.

Slag must be removed using a slag hammer and a wire brush after every pass. If this time-consuming activity is not done, the slag

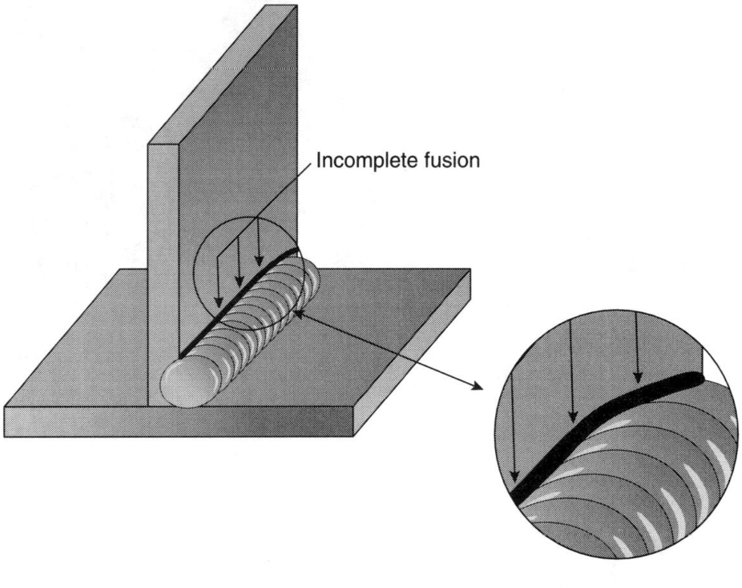

FIGURE 5-8 Incomplete fusion results from an incorrect work angle.

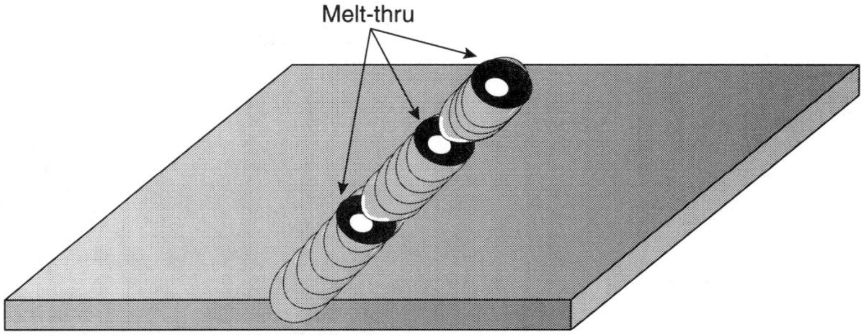

FIGURE 5-9 Melt-thru can be caused by too high a wire speed setting or too slow a travel speed.

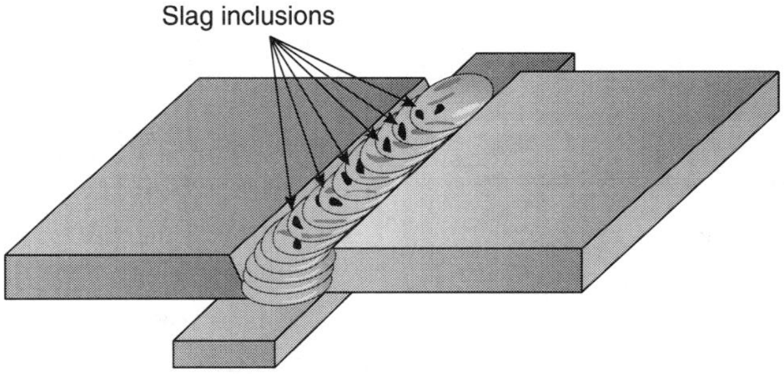

FIGURE 5-10 Slag inclusion can affect weld quality.

can become entrapped within the weld (Figure 5-10). This **slag inclusion** can affect the quality of the weld and possibly cause failure of the joint. Slag inclusions can occur when nonmetallic material residue becomes trapped in the undercut areas of the weld. Slag inclusions can also result from incorrect wire speed or voltage settings and improper welding techniques, including incorrect gun angle or low travel speeds at high wire speed settings.

Slag can become a problem if it begins to flow ahead of the arc (Figure 5-11). This can happen if the welder is not careful; uses too

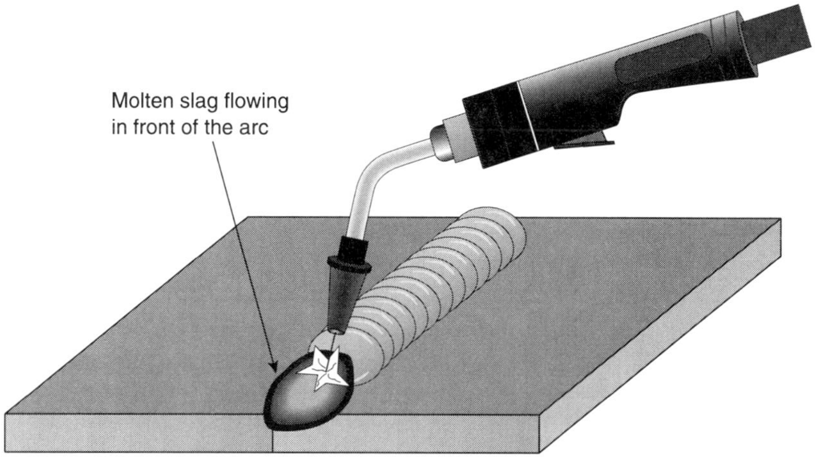

FIGURE 5-11 Molten slag flowing to the front of the arc must be avoided.

slow a travel speed or incorrect gun angle. This problem is more likely to happen when the welder is welding out of position; for example, when welding **downhill,** the effects of gravity can cause the slag to flow down in front of the weld pool.

SAFETY REMINDER
Protect your sight by always wearing safety glasses under the welding helmet to keep hot slag from ending up in your eyes.

Cracking

Cracking of the weld can be a problem which is most likely to occur when welding steels of medium carbon and alloy steels. Welding procedures have to be established for the base metal which involves the selection of the wire electrode. In some applications preheating the base metal to a temperature of approximately 300° to 500° Fahrenheit (maintain 300° to 500° **interpass** temperature as determined by procedure) can help to control the tendency of cracking in the **heat-affected zone (HAZ)** alongside the weld. There is a need to control the temperature during welding because weld metal subject to rapid cooling can change its inner structure, and that of the HAZ alongside the weld (Figure 5-12).

This HAZ alongside the weld, which is never brought to melting temperature, is nonetheless affected by welding. The hardness of the weld will differ from the hardness of the HAZ, which will vary from the unaffected base metal outside the HAZ. A difference in hardness between these three areas is not a problem unless it exceeds specific limitations that can cause cracking. Other undesirable results occur when hardenable steels are hardened and hardened steels are softened in the HAZ. **Underbead cracking** will also occur when hydrogen gas becomes trapped in the weld of hardenable steels (containing over 0.4% carbon) that cools too quickly before the hydrogen can escape.

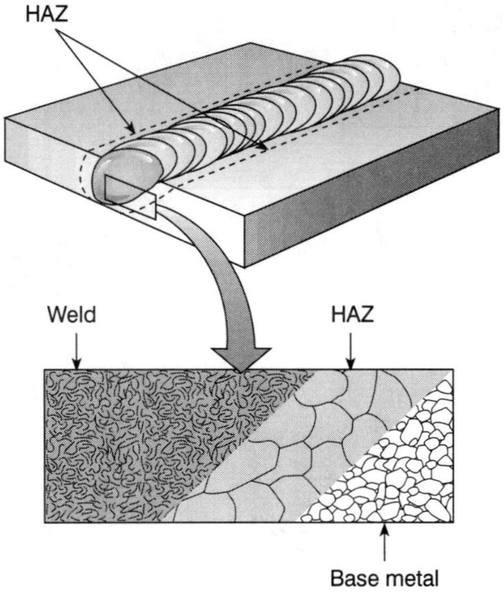

FIGURE 5-12 The steel is coarse grained in the heat-affected zone.

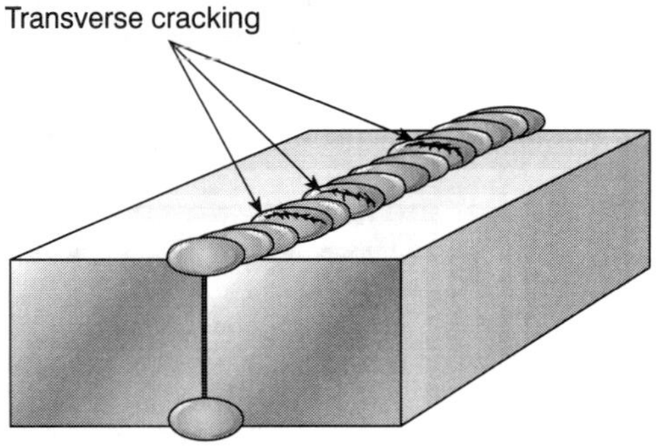

FIGURE 5-13 Transverse cracking is perpendicular to the length of the weld bead.

Preheating most any steel over ½ inch in thickness sometimes works as a remedy to prevent rapid cooling that occurs in thick plate. It is always best to know the type of steel being welded to determine any special welding procedures that must be followed to prevent cracking. Furthermore, be sure to fill each crater at the end of a weld to avoid **crater cracks.** Moving more slowly with a slight sideward motion to produce a larger weld bead and making multiple passes with stringers will also reduce cracking.

Transverse cracking can be a **defect** or **discontinuity** that occurs in the weld from weld toe to weld toe (Figure 5-13). Transverse cracking can result from using the wrong electrode, making too small a weld for the size of the joint, or allowing the weld to cool too quickly. **Longitudinal cracking** can be a defect or discontinuity that occurs in the weld face or the base metal parallel to the weld toes (Figure 5-14). Longitudinal cracking can happen when the joint is unable to move or when it moves too slowly as the weld cools. Choosing a more ductile filler metal, selecting a less rigid joint design, or preheating and slow cooling of the joint will sometimes help.

Cracking is the result when metal under stress is relieved of energy. Welding steels that require special heating and cooling procedures are more susceptible to cracking than low-carbon steels. Welders should know the different cracks brought about

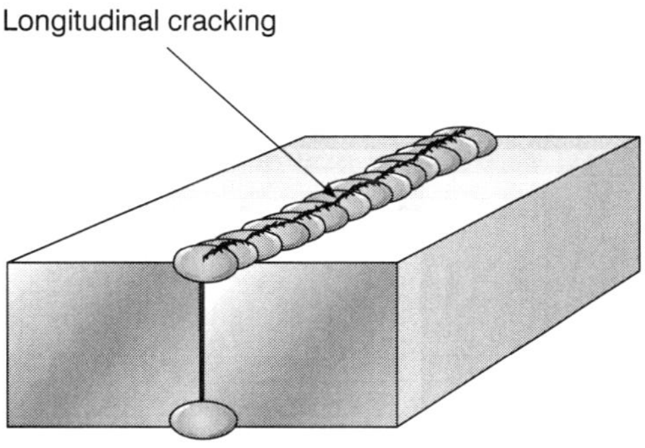

FIGURE 5-14 Longitudinal cracking is parallel to the length of the weld bead.

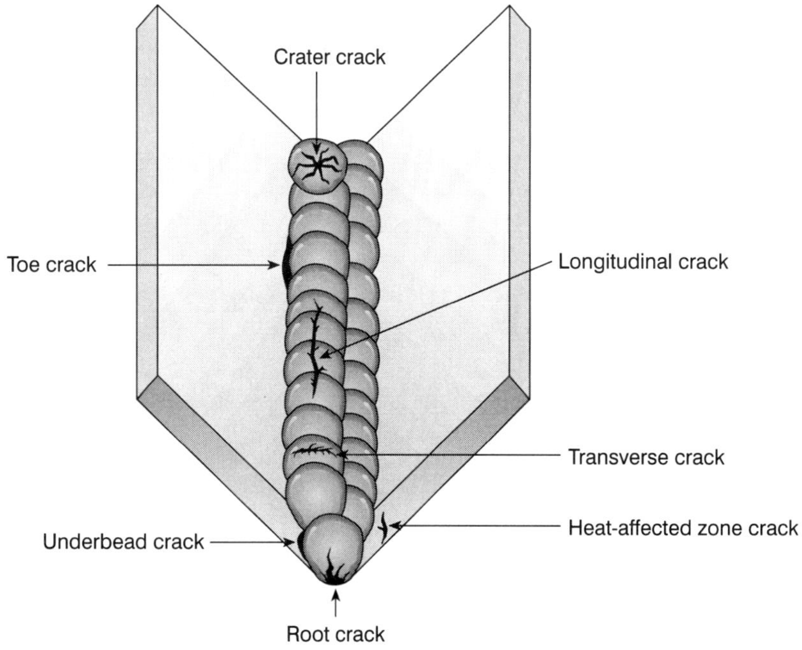

FIGURE 5-15 Several different kinds of cracks can result from welding.

by welding (Figure 5-15), but realize too, the problem of cracking is complex.

Arc Blow

Arc blow is an erratic arc that refuses to go where the welder desires, thus causing spatter. Arc blow is formed by an unbalanced condition that creates a concentration of magnetic fields. When arc blow occurs, the welder may be unable to control where the weld bead is deposited. The bead does not flatten out, and the slag that forms becomes difficult to remove. While the arc may veer left to right during arc blow, normally the arc veers back and forth. The arc veers backward when welding toward the workpiece connection and forward when welding away from the workpiece connection (Figure 5-16).

Arc blow usually is not a problem with AC welding because the current changes direction constantly. Arc blow is a problem for DC and becomes more intense as the amperage is increased. Gas metal and flux cored arc welding use direct current electrode

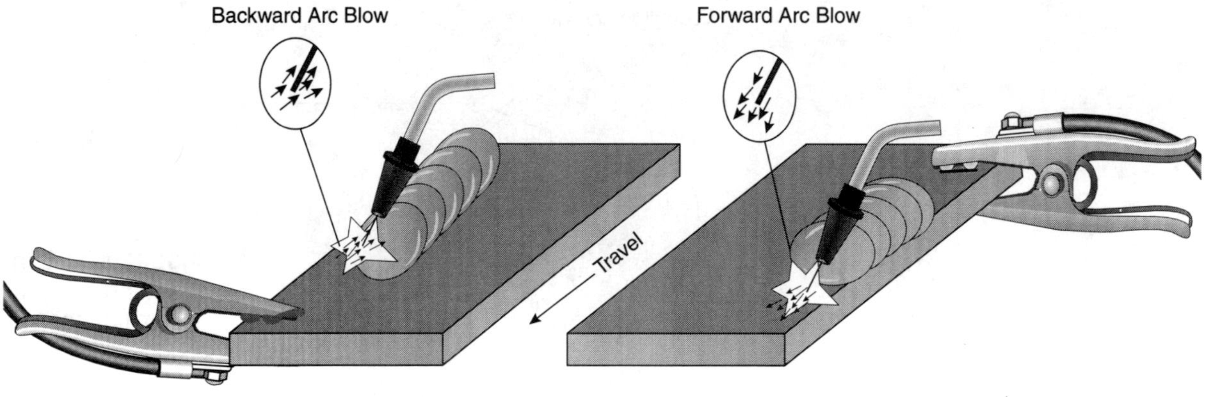

FIGURE 5-16 Arc blow is a common welding problem.

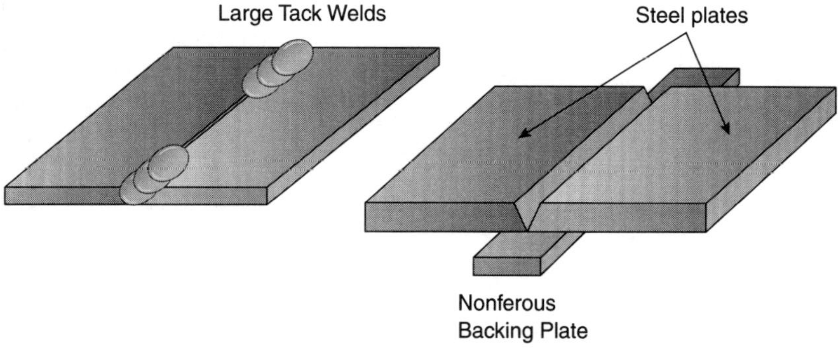

FIGURE 5-17 Two ways of attempting to minimize arc blow.

positive (DCEP) and sometimes direct current electrode negative (DCEN), supplied by the power source. Arc blow can occur in deep groove welds, when welding into or out of corners, or when welding on magnetized metal.

The most common backward arc blow is recognized by excessive spatter; a narrow, high bead; with difficulty removing slag. Forward arc blow decreases in intensity as the distance increases from the workpiece connection. Forward arc blow has a wide but uneven bead, with undercut and incomplete root penetration.

Controlling arc blow. Following are some suggestions for dealing with arc blow. Arc blow is sometimes controllable by switching to AC or changing the placement of the workpiece connection (at the beginning of the weld or at the end of the weld), or by changing the welding direction (from left to right, or vice versa). Try making larger tack welds, hold a shorter arc length, use backing plates on groove welds, or use starting plates to begin the arc or runoff plates to end the arc (Figure 5-17). Instead of placing the workpiece connection on a small project, try clamping the project to a large piece of metal on which the workpiece connection is attached. Ensure good contact of the workpiece connection and remove from the area all metal not being welded. Adjust the gun angle to a position that will overcome arc blow. Finally, the base metal may have to be demagnetized prior to welding.

TABLE 5-1 The troubleshooting guide: a summary of problems for gas metal and flux cored arc welding

Problem	Cause	Remedy
Incomplete root penetration	1. Travel speed too fast	Slow travel speed
	2. Low amperage	Increase wire speed
	3. High voltage	Decrease arc voltage
	4. Electrode extension	Decrease electrode extension
	5. Root opening	Increase root opening
	6. Groove angle	Increase groove angle
	7. Root face	Decrease root face
	8. Position of the arc	Keep arc on front edge of weld pool
Lack of fusion	1. Travel speed too fast	Decrease travel speed
	2. Low amperage	Increase wire speed
	3. High voltage	Decrease arc voltage
	4. Excessive electrode extension	Decrease stickout
	5. Weld pool too large	Increase travel speed
	6. Position of the arc	Keep arc on front edge of weld pool
	7. Base metal preparation	Clean welding surfaces
	8. Groove angle	Increase groove angle
Undercut	1. Travel speed too fast	Decrease travel speed
	2. High voltage	Decrease arc voltage
	3. High amperage	Decrease wire speed
	4. Deficient dwell	Pause longer on side of joint
	5. Gun angle	Change direction for the force of the arc
Porosity	1. Shielding gas flow too low	Increase gas flow
	2. Shielding gas flow too high	Decrease gas flow
	3. Windy or drafty	Barrier protection of welding area
	4. Leaky or restricted gas line	Fix gas line problem
	5. Plugged gun nozzle	Clear spatter from nozzle
	6. Electrode contamination	Use clean, dry wire filler metal
	7. Base metal contamination	Clean base metal before welding
	8. High voltage	Decrease arc voltage
	9. Electrode extension	Decrease stickout
	10. Arc blow	Try suggestions mentioned
Arc melts back to contact tip	1. Voltage too high	Decrease voltage
	2. Wire feed speed too low	Increase wire feed speed
	3. Wire kinked	Fix wire
Melt-thru	1. Travel speed too slow	Increase travel speed
	2. Voltage and amperage too high	Decrease voltage and wire speed
	3. Root opening too large	Reduce the root opening
	4. Electrode extension too short	Increase the stickout
	5. Incorrect gun angle	Change direction for force of the arc
	6. Root face too narrow	Increase root face
Whiskers	1. Pushing wire in front of weld pool	Keep arc on front edge of weld pool
	2. Electrode extension too short	Increase the stickout
	3. Amperage too high	Decrease wire speed
	4. Travel speed too fast	Decrease travel speed
Weld cracks	1. Weld bead too small	Increase size of weld bead
	2. Incorrect wire selection	Match wire to base metal
	3. Base metal temperature too low	Preheat joint adequately
	4. Excessive joint restraint	Allow joint movement during heating and cooling
Excessive spatter	1. Incorrect travel angle	Use backhand technique
	2. Arc voltage too low	Increase arc voltage
	3. Arc blow	Try suggestions mentioned
	4. Nozzle too far away	Bring nozzle closer

REVIEW

1. What causes porosity?
2. What causes undercut?
3. What causes spatter?
4. What causes incomplete fusion?
5. What causes melt-through?
6. What causes slag inclusions?
7. What are the effects of arc blow?

REVIEW QUESTIONS FOR UNIT 5

Multiple Choice

Choose the *best* answer to complete the statement.

1. Weld bead:
   a. Can be weaved.
   b. Is deposited metal.
   c. Can be a stringer.
   d. Is all of the above.
2. Incomplete penetration results when:
   a. Amperage is too high.
   b. Travel speed is too slow.
   c. Wire speed is too low.
   d. Voltage is low.
   e. None of the above.
3. Causes wire to burn back on contact tube:
   a. Long electrode extension
   b. Low amperage with high voltage
   c. High amperage with low voltage
   d. None of the above
4. Will cause undercut:
   a. Wire speed too low
   b. Short electrode extension
   c. Travel speed too fast
   d. None of the above
5. Causes overlap:
   a. Travel speed too fast
   b. Very low wire speed
   c. Arc on trailing side of weld pool
   d. All of the above
   e. None of the above
6. Slag:
   a. Is nonmetallic material residue trapped in the weld.
   b. Protects weld pool from contamination.
   c. Must be removed.
   d. Can become trapped in weld.
   e. Is all of the above.

7. The heat-affected zone (HAZ):
   a. Is heating of base metal.
   b. Is brought to melting temperature.
   c. Is the same hardness as base metal.
   d. Creates a concern with cracking.
   e. Is a. and d. above.
   f. Is a. and c. above.
8. Preheating is:
   a. A special welding procedure.
   b. Used to prevent cracking.
   c. Common on metal over ½-inch thick.
   d. All of the above.
   e. None of the above.
9. To prevent arc blow:
   a. Change placement of workpiece connection.
   b. Provide solid contact of workpiece connection.
   c. Change direction of welding.
   d. Proceed with all of the above.
   e. Proceed with none of the above.

Short Answer

1. Explain the difference between a stringer and a weave weld bead.
2. List one remedy for undercut that does not require adjusting the welding parameters.
3. Explain why is it important to eliminate problems that can occur during welding.
4. List the causes for the formation of craters.
5. Explain the importance of slag removal.
6. Explain how welding downhill can affect the formation of slag.
7. State the likely problem of welding medium carbon and alloy steels.
8. Define *underbead cracking.*
9. Explain the difference between transverse and longitudinal cracking.

ADDITIONAL ACTIVITIES

Matching

Match problem with its causes (hint: problems have more than one cause).

Problem	Causes
Incomplete root penetration	a. use backhand technique
Lack of fusion	b. preheat joint adequately
Undercut	c. travel speed too slow
Porosity	d. arc blow
Melt-thru	e. high amperage
Whiskers	f. weld pool too large
Weld cracks	g. low amperage
Excessive spatter	h. travel speed too fast

Writing Skills

1. In a paragraph, explain arc blow.
2. Briefly describe problems from using too fast a travel speed.

Math

Read each measuring tape and write the appropriate measure on the space provided.

1. ____________

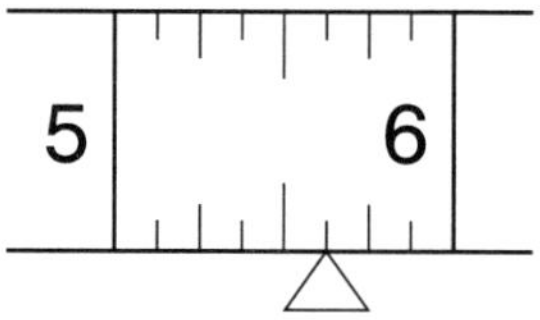

2. ____________

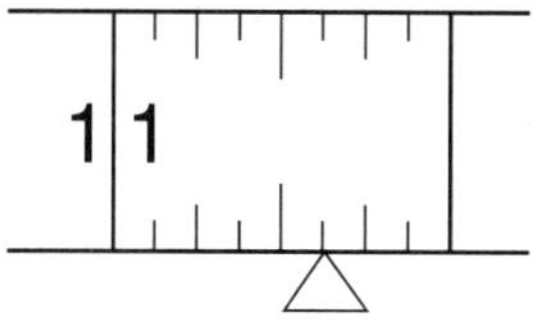

3. ____________

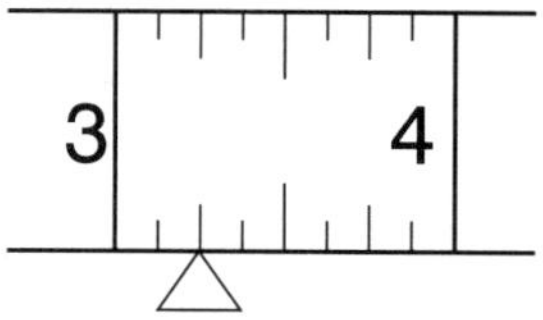

4. ____________

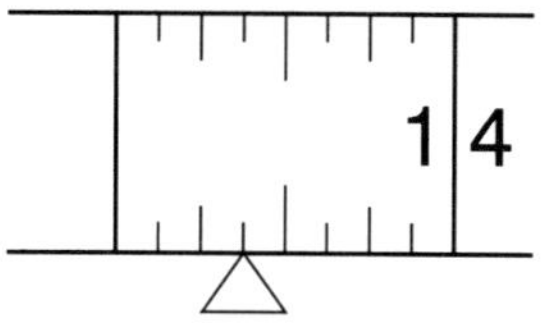

5. ____________

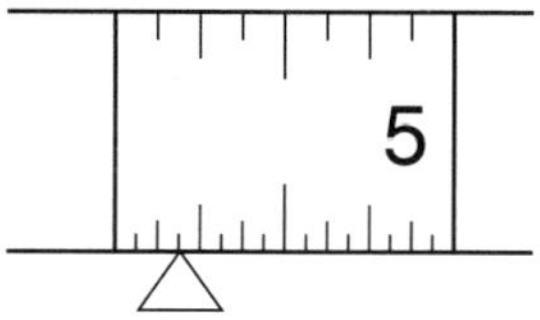

6. ____________

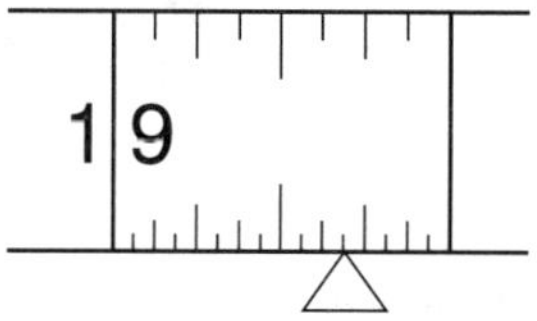

7. ____________

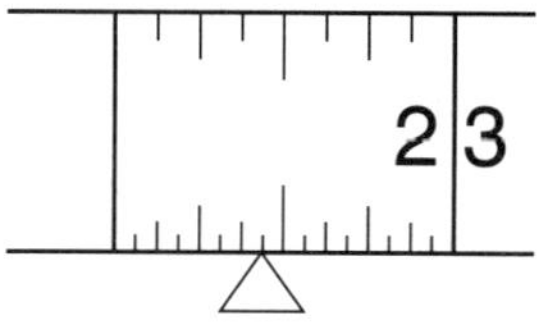

8. ____________

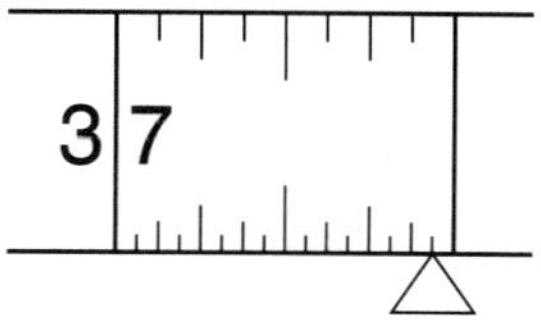

Puzzlers

1. After welding for several minutes along the joint, suddenly the arc begins to throw metal out of the weld. What is a possible problem? What are possible solutions?
2. The wire stops and melts to the contact tube. If the power is still on and the spool full of wire, by troubleshooting the equipment, what could be the problem?
3. The gun becomes very hot. What are some possible causes?

UNIT 6

Welding Exercises

TWO POSITIONS FOR WELDING

Flat Position and Horizontal Position

The American Welding Society (AWS) has established the boundary for the four basic welding positions. This unit will focus on exercises for both the flat position and the horizontal position. The appendix at the back of this book contains exercises for the **overhead position** and for uphill welding in the vertical position.

A position is considered flat if the plane of welding is from 0° to 15° (Figure 6-1). Most welding is done in the flat position because this position allows more metal to be deposited into the joint in a single pass. More metal put into a joint, means more work can be done in the shortest time at the least cost.

Welding in the flat position requires the least amount of skill. The most time and energy learning either GMAW or FCAW also involves the flat position. A beginning welder who invests a little more time to become proficient in the flat position will benefit

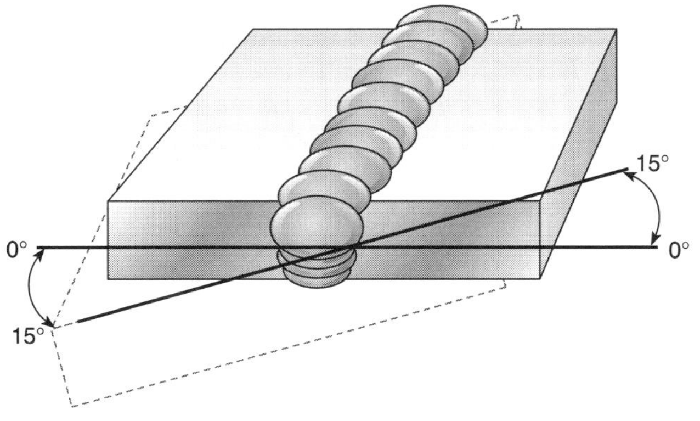

FIGURE 6-1 Plane for the flat position is from 0° to 15°.

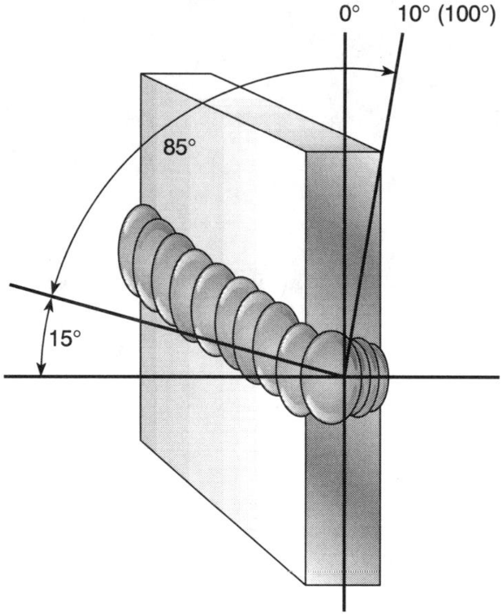

FIGURE 6-2 Plane for the horizontal position.

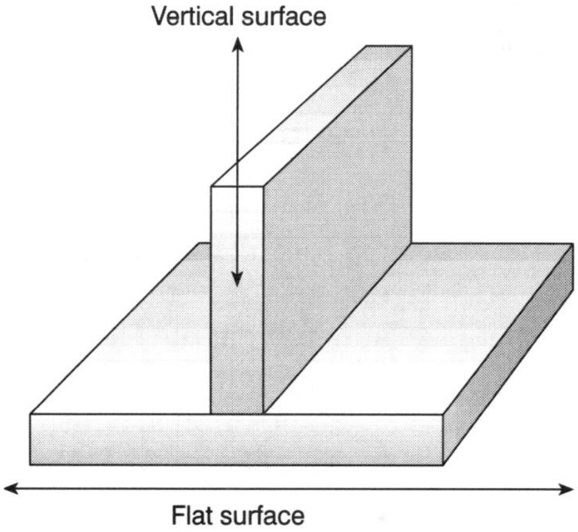

FIGURE 6-3 The tee joint.

when learning to weld in the other positions. Consequently, learning to weld in the horizontal position will be easier.

A position is considered horizontal if the plane of welding is from 15° to 100° (Figure 6-2). Being able to weld in the horizontal position will expand a welder's capability to weld the joint designs on many projects. The tee joint is the best example of welding in the horizontal position (Figure 6-3). The weld is along a joint with surfaces both flat and vertical.

The following exercises can be completed using 16-gauge to ½-inch-thick steel. The suggested sizes are 16-gauge sheet × 3″ × 8″; and plates ¼″ × 3″ × 8″, ⅜″ × 6″ × 8″, and ½″ × 3″ × 8″. Steel ⅜ inch thick is recommended for some exercises, but all the exercises can be accomplished using ⅜-inch material. Six pieces of 6″ × 8″ plate and twelve pieces of 3″ × 8″ plate are enough to start working on the exercises presented in this unit. Please keep in mind that the dimensions provided here are suggestions. Many shops have to depend on the material that is available at any given time. Note, several of the exercises are ideal for small 110-volt power sources that operate off household current.

REVIEW

1. List some differences between the flat position and the horizontal position.
2. Sketch a lap joint in a three-dimensional view.

LAB EXERCISE ONE

Short Circuiting Transfer (GMAW-S): Laying a Bead

During the setup for welding, remember that the voltage and wire speed adjustments work together. If the wire speed adjustment is too high for the voltage, the wire will not melt properly into the weld pool, and excessive buildup of filler metal will result. But if the wire speed adjustment is too low for the voltage, the wire will burn back into the contact tube. To save the contact tube from becoming fused to the wire during welding, begin with a high wire speed and low voltage adjustments. Lay several beads at different settings, and observe how each weld bead looks (Figure 6-4).

Equipment and Material

Safety glasses
Helmet
Earplugs
Grinder
Shielding gas: carbon dioxide or 75% argon to 25% carbon dioxide
Welding gloves
Protective clothing
Sidecutter
1 plate ⅜-inch steel, 6″ × 8″
Spool wire 0.035 (ER70S-3)

Welding Parameters

Amperes: 200 to 225
Voltage: 22 to 24
Electrode extension: ¼ to ⅜ inch

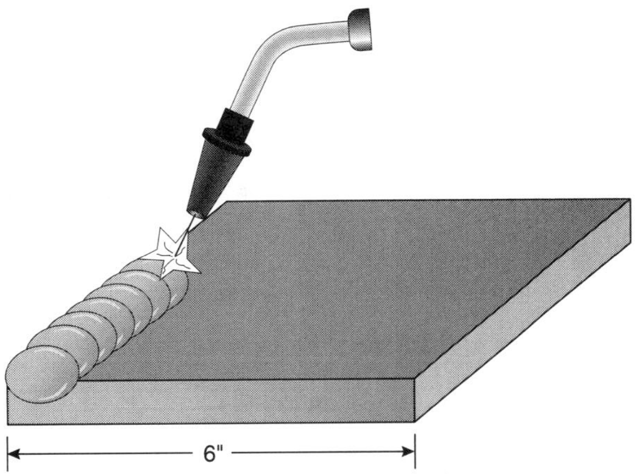

FIGURE 6-4 Lay the first bead along the edge of the plate.

Directions

1. Clean the surface of the plate with a grinder.
2. Make sure the power is off before plugging in the power source.
3. Screw out the pressure regulator (if not attached to a flowmeter) before opening the gas cylinder valve, standing to one side. Follow safe practices when either attaching or replacing a cylinder. Move the cylinder only with the valve cap in place, and secure the cylinder so it cannot fall.
4. Turn on the power source and set with 22 to 24 volts. Adjust the gas flow from 15 to 25 cubic feet per hour (cfh).
5. Set the wire feed at 50% and adjust it after doing a test weld bead.
6. Make several weld beads.
7. Test the welds by visual inspection.
8. Check with the instructor for evaluation, if necessary, before proceeding to the next exercise.

LAB EXERCISE TWO

Short Circuiting Transfer (GMAW-S): Padding Plate in Flat Position

Surfacing Weld

Vision is limited under the helmet. Lay the first weld bead along the back edge of the plate (Figure 6-5). Use the edge as a guide

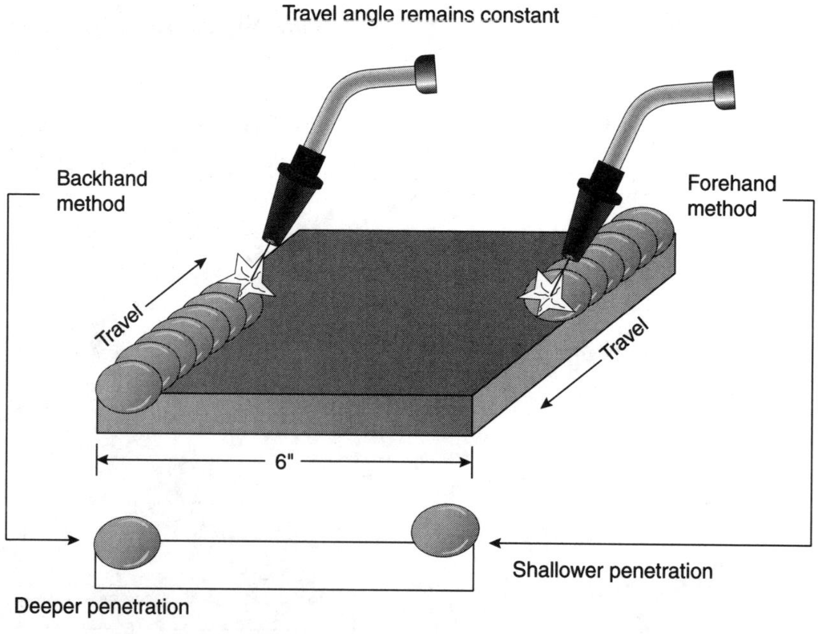

FIGURE 6-5 Making a paddling plate using both forehand and backhand methods.

for keeping that first bead straight. Make one layer using the backhand method, then run one layer of weld beads using the forehand method. Work on making weld beads of consistent size and concentrate on maintaining proper travel speed and gun angle. Once again, do not be afraid to make adjustments on the controls to examine the results of different settings. This exercise is an example of the kind of welding used to build up worn parts.

Equipment and Material

Safety glasses
Helmet
Earplugs
Grinder
Shielding gas: carbon dioxide or 75% argon to 25% carbon dioxide
Welding gloves
Protective clothing
Sidecutter
1 plate ⅜-inch steel, 6″ × 8″
Spool wire 0.035 (ER70S-3)

Welding Parameters

Amperes: 200 to 225
Voltage: 22 to 24
Electrode extension: ¼ to ⅜ inch

Directions

1. Clean the surfaces of base metal with the grinder.
2. Make sure the power is off before plugging in the power source. Switch on power.
3. Adjust the gas flow from 20 to 25 cfh.
4. Fill the plate completely with lapping weld beads. Pause to fill craters at the end of each weld bead.
5. The ripples provide a pattern that should be rounded. Ripples that come to a point indicate a travel speed that is too fast. A high, narrow weld bead suggests the arc length is too short, possibly from using too high a wire speed setting or too low a voltage setting.
6. If the weld bead flattens with uneven ripple pattern and spatter, the voltage setting could be too high.
7. Make straight weld beads with even ripples and consistent **lapping** of one weld bead to another.
8. Watch to maintain a 45° travel angle.
9. Complete at least two layers of weld beads, practicing both backhand and forehand methods.
10. Test the weld by visual inspection.
11. Check with the instructor for evaluation, if necessary, before proceeding to the next exercise.

Short Circuiting Transfer (GMAW-S): Corner Joint in Flat Position

Fillet Weld

Exercise three requires consistent welding with complete fusion of a fillet weld into the joint without undercutting the base metal. Occasionally examine the nozzle for spatter, which can affect gas flow. An antispatter spray or gel can be used. The setup for this exercise is illustrated in Figure 6-6.

The tack weld, introduced with this exercise, is used when two pieces are to be joined. Tack welds are short, temporary weld beads used to hold a weldment in position to complete welding. Tack welds are important in any fabrication project. All parts essential to the weldment must be tack welded first.

Equipment and Material

Safety glasses
Helmet
Earplugs
Grinder
Welding gloves

Pieces of plate ⅜-inch steel, 3″ × 8″ (2)
Sidecutter
Protective clothing

Spool wire 0.035 (ER70S-3)
Padding plate (scrap)

Shielding gas: carbon dioxide or 75% argon to 25% carbon dioxide

Welding Parameters

Amperes: 200 to 225
Voltage: 22 to 24
Electrode extension: ¼ to ⅜ inch

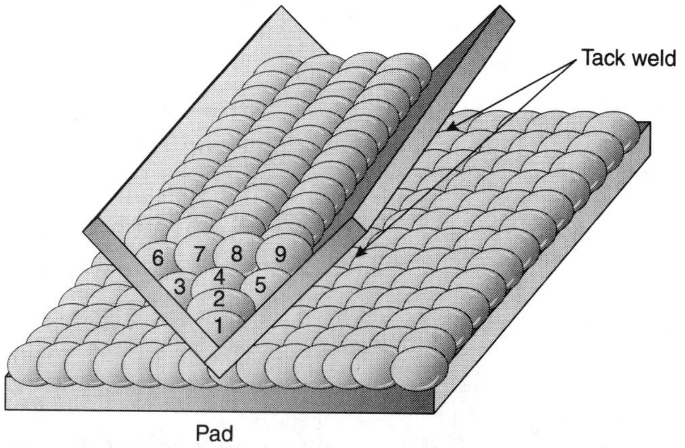

FIGURE 6-6 Lay each weld bead in a sequence.

Directions

1. Clean the surfaces of the base metal with the grinder.
2. Make sure the power is off before plugging in the power source. Switch on power.
3. Adjust the gas flow from 20 to 25 cfh.
4. Set the voltage and set the wire speed at 50% and adjust by making test runs. Observe voltmeter and ammeter during welding to adjust the welding parameters.
5. Tack weld together the two pieces of plate, and then tack weld them to the scrap padding plate.
6. Lay eight to twelve weld beads with a slight sideways motion. Practice both forehand and backhand methods.
7. Observe the lapping of one weld bead tying in to another. There should be no valleys or grooves between weld beads.
8. Look for undercut of weld beads along the base metal. If the voltage and wire speed are set correctly, check the angle of the gun. Cool plate when necessary.
9. Test by visual inspection for weld bead consistency and lapping, avoiding undercut.
10. Check with the instructor for evaluation, if necessary, before proceeding to the next exercise.

LAB EXERCISE FOUR

Short Circuiting Transfer (GMAW-S): Lap Joint Horizontal Position

Fillet Weld

The lap joint tests an ability to make weld beads of consistent size in a limited space. There must be enough lapping to keep the legs of the fillet weld equal to the thickness of the base metal (Figure 6-7).

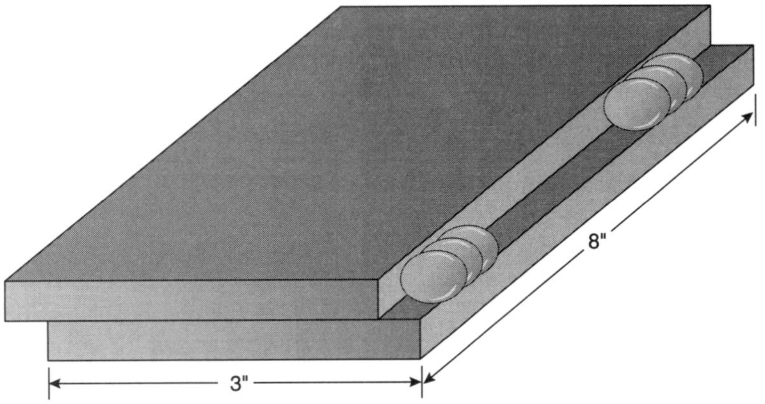

FIGURE 6-7 Tack weld first before laying weld beads.

Use the edge as a guide while making each pass. Watch the weld pool and maintain a fast enough travel speed to keep the top piece from melting away.

Equipment and Material

Safety glasses	Welding gloves	Sidecutter
Helmet	Pieces of plate	Protective clothing
Earplugs	¼ to ⅜-inch steel, 3″ × 8″ (2)	Spool wire 0.035 (ER70S-3)
Grinder		

Shielding gas: carbon dioxide or 75% argon to 25% carbon dioxide

Welding Parameters

Amperes: 140 to 250
Voltage: 19 to 24
Electrode extension: ¼ to ⅜ inch

Directions

1. Clean the surfaces of the base metal with the grinder.
2. Make sure the power is off before plugging in the power source. Switch on power.
3. Adjust the gas flow from 20 to 25 cfh.
4. Set the voltage and set the wire speed; adjust by making test welds. Observe voltmeter and ammeter during welding to adjust welding parameters.
5. Use the backhand method.
6. Test by visual inspection. The leg sizes of the weld should equal the thickness of the base metal.
7. Check with the instructor for evaluation, if necessary, before proceeding to the next exercise.

LAB EXERCISE FIVE

Short Circuiting Transfer (GMAW-S): Lap Joint on Sheet Steel in Horizontal Position

Fillet Weld

Make several tack welds to help minimize the problem of distortion when welding thin metal (Figure 6-8). Tack welds are a form of **intermittent welding.** Begin by laying one continuous weld bead to develop skill in welding thin metal. The key to welding a joint of this type is to set up a close fitting joint. If necessary, use

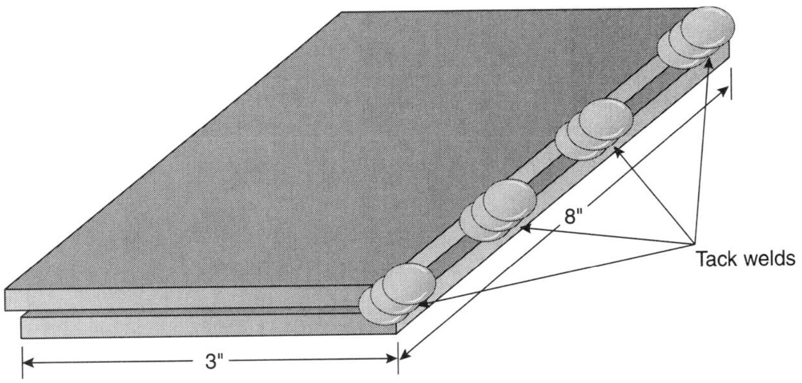

FIGURE 6-8 Distortion is always a problem when using sheet steel.

a slight weave motion for fusing together the edges of both pieces in the joint.

Equipment and Material

Safety glasses
Helmet
Earplugs
Grinder
Welding gloves
Pieces of 16-gauge steel, 3″ × 8″ (2)
Sidecutter
Protective clothing
Spool wire 0.035 (ER70S-3)
Shielding gas: carbon dioxide or 75% argon to 25% carbon dioxide

Welding Parameters

Amperes: 80 to 110
Voltage: 18 to 20
Electrode extension: ¼ to ⅜ inch

Directions

1. Clean the surfaces of the base metal with the grinder.
2. Make sure the power is off before plugging in the power source. Switch on power.
3. Adjust the gas flow from 15 to 25 cfh.
4. Set the voltage and set the wire speed; adjust by making test welds. Observe voltmeter and ammeter during welding to adjust welding parameters.
5. Use both forehand and backhand methods to establish which method works best. Remember that the backhand method concentrates more heat at the joint root.
6. Make a visual inspection. Look for consistent size in the formation of the weld bead without undercut or porosity.
7. Check with the instructor for evaluation, if necessary, before proceeding to the next exercise.

Short Circuiting Transfer (GMAW-S): Butt Joint on Sheet Steel in Flat Position

Square-Groove Weld

Distortion is a serious problem when using a butt joint of thin metal. If a box and pan brake is available, put a small bend on the edge of very thin metal (thinner than 16 gauge) for less distortion in an edge flange joint. The bend in the flange melts with the filler metal into the weld. If the welding parameters are set correctly, the travel speed is important for complete joint penetration. These exercises on thin metal can be more difficult than welding on actual projects. Weldments usually contain more mass, resulting in less distortion (Figure 6-9).

Equipment and Material

Safety glasses
Helmet
Earplugs
Grinder
Welding gloves
Shielding gas: carbon dioxide or 75% argon to 25% carbon dioxide
Protective clothing
Sidecutter
Pieces of 16-gauge steel, 3″ × 8″ (2)
Spool wire 0.035 (ER70S-3)
C-clamp fixture

Welding Parameters

Amperes: 80 to 110
Voltage: 18 to 20
Electrode extension: ¼ to ⅜ inch

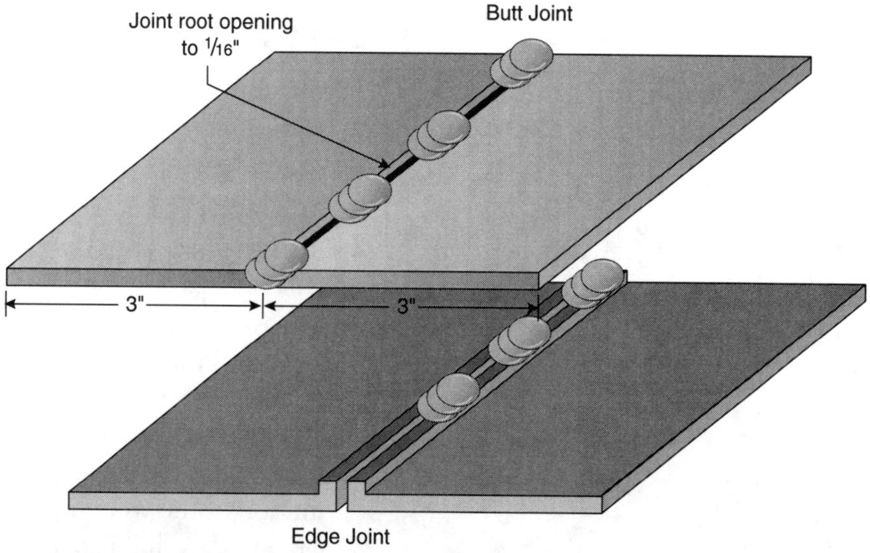

FIGURE 6-9 Maintain a joint root opening to aid in achieving 100% complete joint penetration.

Directions

1. Clean the surfaces of the base metal with the grinder.
2. Make sure the power is off before plugging in the power source. Switch on power.
3. Adjust the gas flow from 15 to 25 cfh.
4. Set the voltage and set the wire speed; adjust by making test welds. Observe voltmeter and ammeter during welding to adjust welding parameters.
5. Tack weld the pieces of the joint together. Place the root opening off the table, using a C-clamp fixture, fire brick, or other device.
6. Use the forehand technique, concentrating on **complete joint penetration.**
7. Make a visual inspection. Look for complete joint penetration on the underside of the joint with no porosity.
8. Check with the instructor for evaluation, if necessary, before proceeding to the next exercise.

LAB EXERCISE SEVEN

Short Circuiting Transfer (GMAW-S): Butt Joint on Sheet Steel in Horizontal Position

Square-Groove Weld

Exercise Seven is similar to Exercise Six, except the joint is welded in the horizontal position without using a flange to limit distortion (Figure 6-10). Keep the gun at a 90° work angle, and move along the joint at a constant pace to avoid melt-thru. Let travel speed be determined by what is necessary to achieve complete joint penetration. If penetration is incomplete, reduce the electrode extension.

Equipment and Material

Safety glasses
Helmet
Earplugs
Grinder
Welding gloves
Shielding gas: carbon dioxide or 75% argon to 25% carbon dioxide
Protective clothing
Sidecutter
Pieces of 16-gauge steel, 3″ × 8″ (2)
Spool wire 0.035 (ER70S-3)
C-clamp fixture

Welding Parameters

Amperes: 80 to 110
Voltage: 18 to 20
Electrode extension: ¼ to ⅜ inch

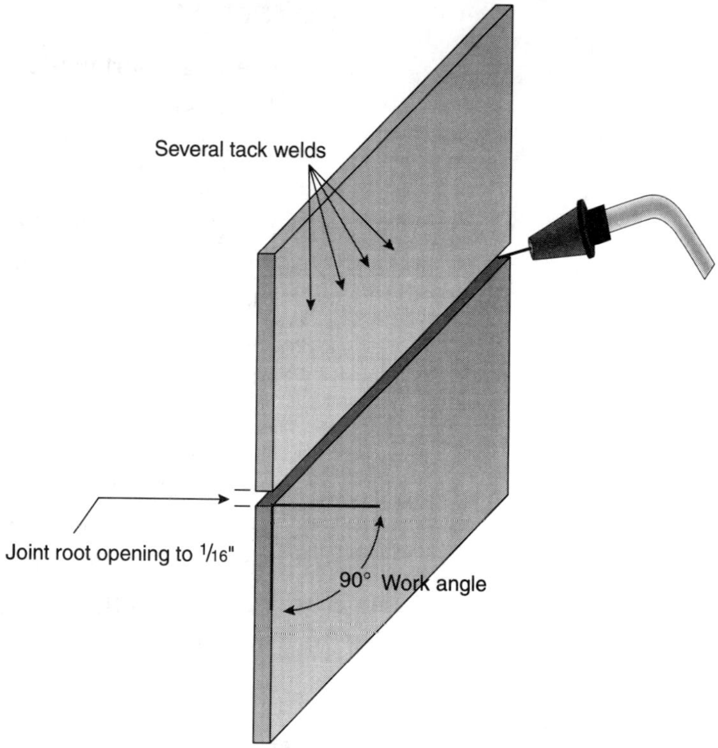

FIGURE 6-10 Pay attention to work angle and travel speed for complete joint penetration.

Directions

1. Clean the surfaces of the base metal with the grinder.
2. Make sure the power is off before plugging in the power source. Switch on power.
3. Adjust the gas flow from 15 to 25 cfh.
4. Set the voltage and set the wire speed; adjust by making test welds. Observe voltmeter and ammeter during welding to adjust welding parameters.
5. Use the forehand method, concentrating on complete joint penetration.
6. Make a visual inspection. Look for complete joint penetration and no porosity.
7. Check with the instructor for evaluation, if necessary, before proceeding to the next exercise.

LAB EXERCISE EIGHT

Short Circuiting Transfer (GMAW-S): Lap Joint on Sheet Steel in Vertical Position

Fillet Weld

Exercise Eight is designed for learning the downhill welding technique (Figure 6-11). Downhill is useful in welding very thin metal

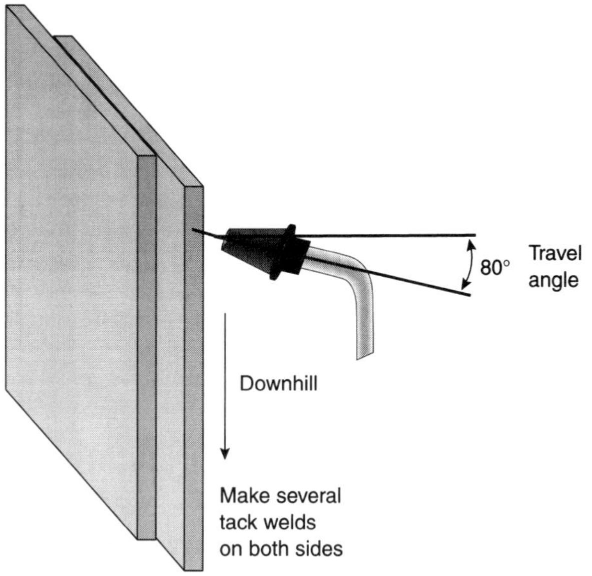

FIGURE 6-11 Avoid traveling so slow as to let the weld pool drop below the arc.

when melt-thru can be a problem. Knowing how to downhill weld is a useful skill, forcing the welder to maintain a faster travel speed than usual. The faster travel speed—caused by gravity—generates less heat.

Equipment and Material

Safety glasses
Helmet
Earplugs
Grinder
Welding gloves
Protective clothing
Sidecutter
Pieces of 16-gauge steel, 3″ × 8″ (2)
Spool wire 0.035 (ER70S-3)
C-clamp fixture
Shielding gas: carbon dioxide or 75% argon to 25% carbon dioxide

Welding Parameters

Amperes: 80 to 110
Voltage: 18 to 20
Electrode extension: ¼ to ⅜ inch

Directions

1. Clean the surfaces of the base metal with the grinder.
2. Make sure the power is off before plugging in the power source. Switch on power.
3. Adjust the gas flow from 15 to 25 cfh.
4. Set the voltage and set the wire speed; adjust by making test welds. Observe voltmeter and ammeter during welding to adjust welding parameters.

5. Tack weld in several places and position in fixture.
6. Keep the wire electrode in the crack of the inside corner where the two pieces meet.
7. Make a visual inspection. Look for a weld bead of consistent size.
8. Check with the instructor for evaluation, if necessary, before proceeding to the next exercise.

LAB EXERCISE NINE

Short Circuiting Transfer (GMAW-S): Butt Joint on Sheet Steel in Vertical Position

Square-Groove Weld

Controlling the size of the weld pool is important for making weld beads of consistent size (Figure 6-12). Gravity helps to make downhill welding easier. The formation of a key hole, which is the melting back of the edge on each side of the joint, indicates that the weld bead is achieving complete joint penetration. Stay in the root opening and keep the electrode on the leading edge of the weld pool.

Equipment and Material

Safety glasses
Helmet
Earplugs
Grinder
Welding gloves
Protective clothing

Sidecutter
Pieces of 16-gauge steel, 3″ × 8″ (2)
Spool wire 0.035 (ER70S-3)
C-clamp

Shielding gas: carbon fixture dioxide or 75% argon to 25% carbon dioxide

Welding Parameters

Amperes: 80 to 110
Voltage: 18 to 20
Electrode extension: ¼ to ⅜ inch

Directions

1. Clean the surfaces of the base metal with the grinder.
2. Make sure the power is off before plugging in the power source. Switch on power.
3. Adjust the gas flow from 15 to 25 cfh.
4. Set the voltage and set the wire speed; adjust by making test welds. Observe voltmeter and ammeter during welding to adjust welding parameters.

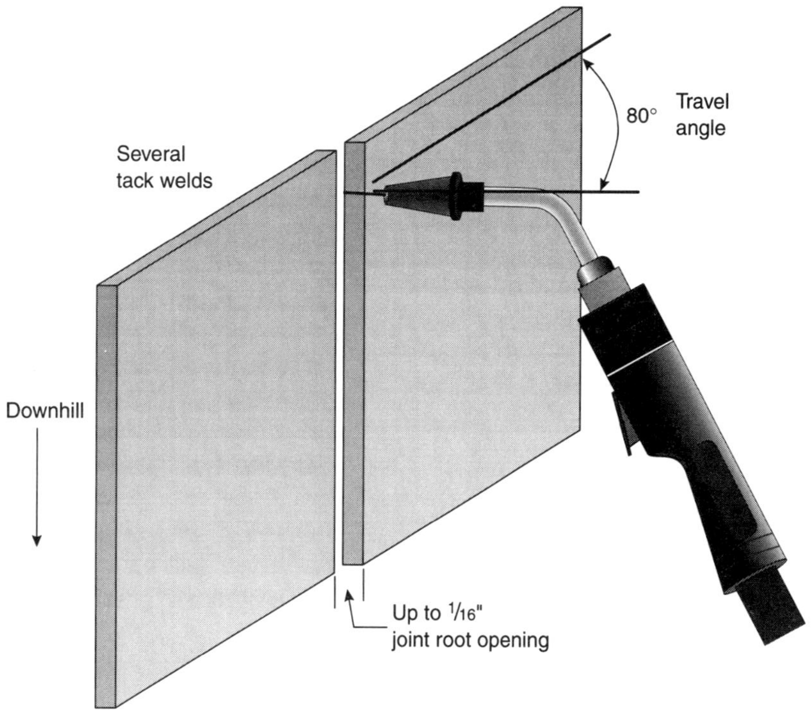

FIGURE 6-12 Be sure to melt the edge of each piece.

5. Tack weld in several places. When making the fit up, the **joint root opening** should equal the diameter of the wire, up to 1/16 inch.
6. Maintain the proper travel angle to complete the weld bead.
7. Make a visual inspection. Look for a straight weld bead of consistent size; turn plate over to examine for complete joint penetration.
8. Check with the instructor for evaluation, if necessary, before proceeding to the next exercise.

LAB EXERCISE TEN

Spray-Arc Transfer: Corner Joint in Flat Position Fillet Weld

Spray-arc transfer occurs when using a shielding gas mixture (suggested 98% argon to 2% oxygen) which differs from short circuiting transfer, and is recommended for carbon steels. The higher heat input necessary for spray-arc transfer requires that the weldment be cooled down on occasion. This method of transfer produces a high rate of deposition with deep penetration into the plate. Do not move so quickly that the arc leaves the weld pool, but fast enough to avoid excessive buildup (Figure 6-13).

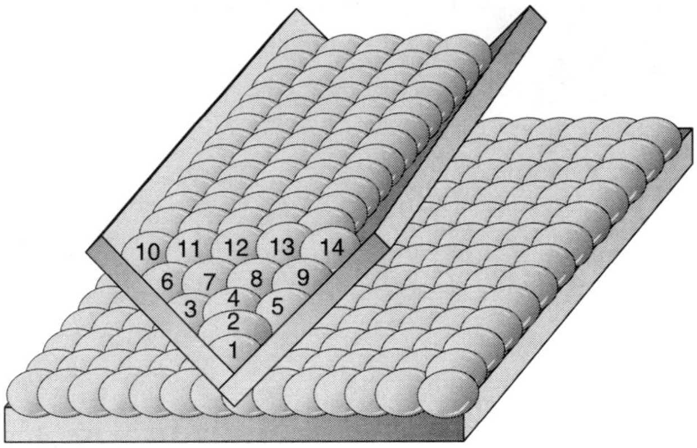

FIGURE 6-13 Fillet weld in the flat position.

Equipment and Material

- Safety glasses
- Helmet
- Earplugs
- Grinder
- Welding gloves
- Protective clothing
- Sidecutter
- Pieces of $\frac{3}{8}$-inch steel, 3″ × 8″ (2)
- Spool wire 0.045 (ER70S-3)
- Shielding gas: 98% argon to 2% oxygen
- Padding plate (scrap)

Welding Parameters

Amperes: 160 to 340

Voltage: 23 to 30

Electrode extension: $\frac{3}{8}$ to $\frac{3}{4}$ inch

Directions

1. Clean the surfaces of the base metal with the grinder.
2. Make sure the power is off before plugging in the power source. Switch on power.
3. Adjust the gas flow from 40 to 50 cfh.
4. Set the voltage and set the wire speed; adjust by making test welds. Observe voltmeter and ammeter during welding to adjust welding parameters.
5. If porosity develops immediately, examine the nozzle for restricted gas flow.
6. Cool the steel as it overheats.
7. Lay 10 to 15 weld beads and examine them for consistency of size. The weld beads should tie together.
8. Make a visual inspection. Look for porosity and undercut.
9. Check with the instructor for evaluation, if necessary, before proceeding to the next exercise.

Spray-Arc Transfer: Tee Joint in Horizontal Position Fillet Weld

The tee joint is popular in the design of weldments. In this exercise the work angle of the gun changes with each weld bead. The first weld bead requires a 45° work angle. The work angle for the second (60° work angle) and the third (30° work angle) weld beads changes also. There has to be enough lapping of the second and third weld beads so that the weld leg sizes are equal. Position the gun at a 45° work angle, with a 15° to 30° travel angle from perpendicular (Figure 6-14).

Equipment and Material

Safety glasses	Protective clothing	Shielding gas: 98% argon to 2% oxygen
Helmet	Sidecutter	Padding plate (scrap)
Earplugs	Pieces of ⅜-inch steel, 3″ × 8″ (2)	
Grinder		
Welding gloves	Spool wire 0.045 (ER70S-3)	

Welding Parameters

Amperes: 160 to 340

Voltage: 23 to 30

Electrode extension: ⅜ to ¾ inch

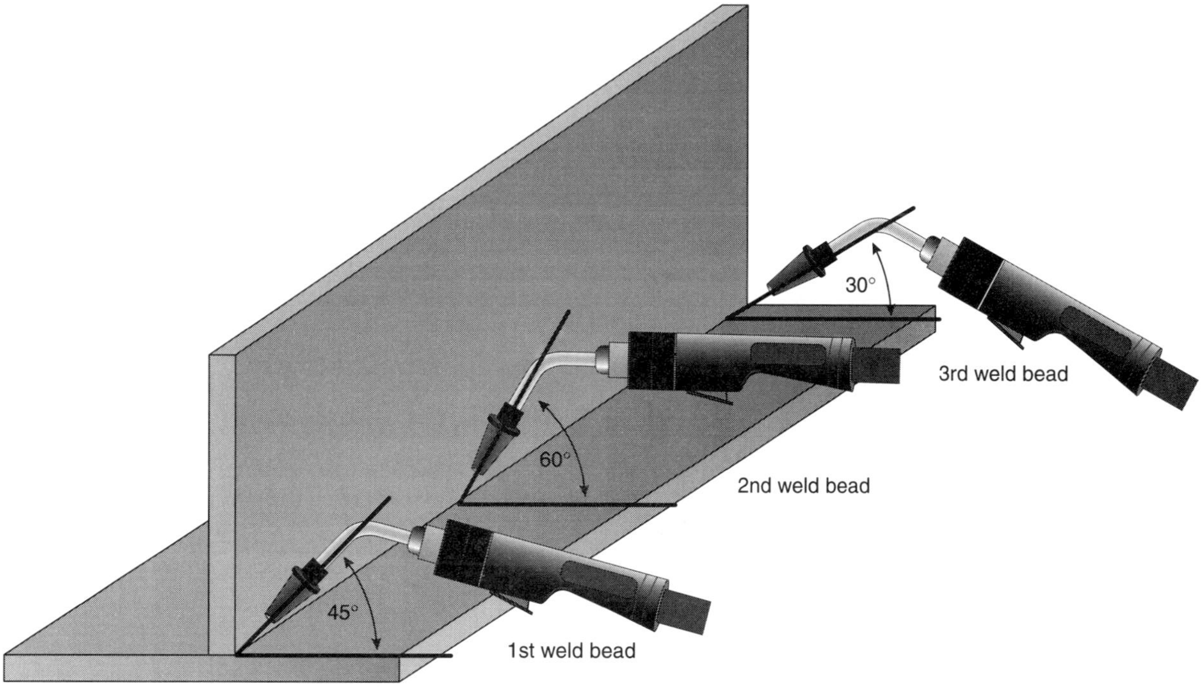

FIGURE 6-14 Work angle changes slightly with each weld bead.

Directions

1. Clean the surfaces of the base metal with the grinder.
2. Make sure the power is off before plugging in the power source. Switch on power.
3. Adjust the gas flow from 40 to 50 cfh.
4. Set the voltage and set the wire speed; adjust by making test welds. Observe voltmeter and ammeter during welding to adjust the welding parameters.
5. Set the parameters for welding on a piece of scrap steel.
6. Cool the joint between passes when overheating occurs.
7. Lay at least three to six weld beads on each side.
8. Make a visual inspection. Look for porosity and undercut.
9. Check with the instructor for evaluation, if necessary, before proceeding to the next exercise.

LAB EXERCISE TWELVE

Spray-Arc Transfer: Butt Joint*

Exercise Twelve is designed for gas shielding but can be completed using FCAW-S; evaluate with a destructive test. Note that a backing strip is used in this exercise. Fit up is important, so be sure that the dimensions are accurate. Go over the tack welds when completing the first pass. Stay in the root opening and keep the electrode on the leading edge of the weld pool (Figures 6-15a and b).

Equipment and Material

Safety glasses
Helmet
Earplugs
Grinder
Welding gloves
Protective clothing
Sidecutter

Pieces of ⅜-inch A36 steel, 3″ × 7″ (2)

Spool wire 0.045 (ER70S-3)

Cutting Machine set at 22.5°

2″ × 10″ **backing** (1 piece)

Shielding gas: 98% argon to 2% oxygen

Strong backs (2)

Welding Parameters

Amperes: 160 to 340

Voltage: 23 to 30

Electrode extension: ⅜ to ¾ inch

---

*Meets AWS structural V-groove weld with backing in flat position—AWS Welding Code D1.1

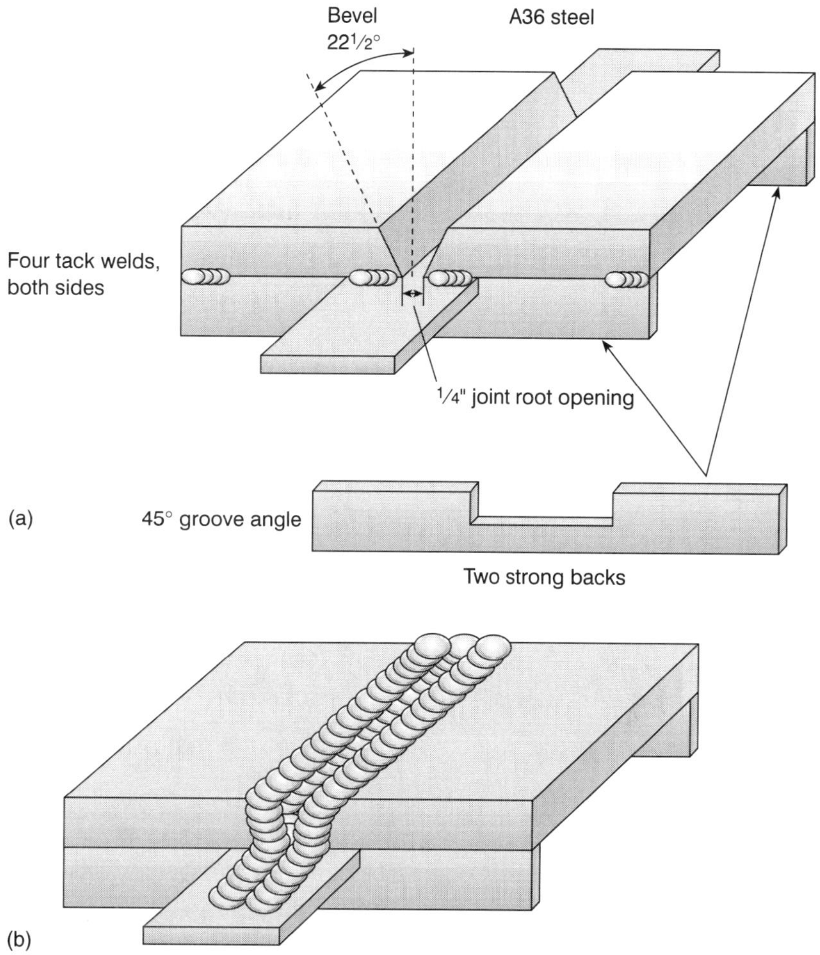

FIGURE 6-15 (a) Preparation of V-groove weld with backing in flat position. (b) Complete V-groove weld in flat position.

Directions

1. Clean the surfaces of the base metal with the grinder.
2. Make sure the power is off before plugging in the power source. Switch on power.
3. Adjust the gas flow from 40 to 50 cfh.
4. Set the voltage and set the wire speed; adjust by making test welds. Observe voltmeter and ammeter during welding to adjust welding parameters.
5. Tack weld the two pieces to the backing plate. Set a ¼-inch joint root opening. Tack weld the strong backs in four places.
6. Use a sideward motion, making sure there is complete fusion on the joint root pass.
7. Make a visual inspection. Look for complete fill within the joint.
8. Check with the instructor for evaluation, if necessary, before proceeding to the next exercise.
9. Pass the guided bend test before proceeding to the next exercise.

LAB EXERCISE THIRTEEN

Flux Cored Arc Welding (FCAW-G): Corner Joint in Flat Position

Fillet Weld

The purpose of Exercise Thirteen is to introduce the FCAW process. Although this exercise is designed for gas shielding, it can be completed using a self-shielding electrode wire. The greater the thickness of the steel plate, the larger the diameter of electrode wire that can be used (Figure 6-16).

Equipment and Material

- Safety glasses
- Helmet
- Earplugs
- Grinder
- Welding gloves
- Protective clothing
- Sidecutter
- Slag hammer
- Wire brush
- Padding plate (scrap)
- Pieces of ⅜-inch steel, 3″ × 7″ (2)
- Spool wire 0.045 (E71T-1)
- Shielding gas: carbon dioxide

Welding Parameters

Amperes: 175 to 200

Voltage: 21 to 22

Electrode extension: ⅜ to ¾ inch

Directions

1. Clean the surfaces of the base metal with the grinder.
2. Make sure the power is off before plugging in the power source. Switch on power.

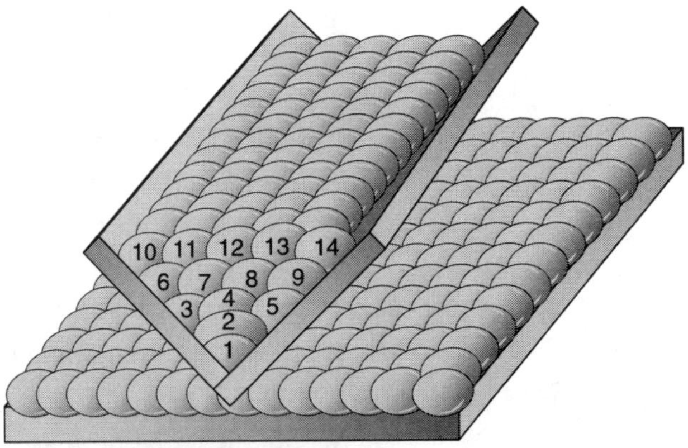

Remove slag between each weld bead

FIGURE 6-16 Keep head out of smoke stream while completing this exercise.

3. Adjust the gas flow from 35 cfh.
4. Set the voltage and set the wire speed; adjust by making test welds. Observe voltmeter and ammeter during welding to adjust welding parameters.
5. Maintain electrode extension from ½ to ¾ inch. To eliminate stubbing (which can occur when the wire is driven through the weld pool) increase the voltage or decrease the wire speed.
6. Cool when necessary. Remove all slag before laying the next pass. Lay several passes.
7. Make a visual inspection. Look for porosity and undercut.
8. Check with the instructor for evaluation, if necessary, before proceeding to the next exercise.

LAB EXERCISE FOURTEEN

Flux Cored Arc Welding (FCAW-G): Butt Joint in Flat Position

Square-Groove Weld with Backing

Exercise Fourteen is designed for gas shielding, but can be completed using a self-shielding electrode wire. The greater the thickness of the steel plate, the larger the diameter of electrode wire that can be used. Although ⅜-inch plate is suggested, thicker plate can be used (Figure 6-17). Use a slight weave motion to wash the edges of both pieces of the joint with the weld pool.

Equipment and Material

Safety glasses	Sidecutter	Shielding gas: carbon dioxide
Helmet	Wire brush	Pieces of ⅜-inch A36 steel, 3″ × 7″ (2)
Earplugs	Slag hammer	⅜-inch steel plate, 2″ × 8″ (backing)
Grinder	Spool wire 0.045 (E71T-1)	
Welding gloves	Strong backs(2)	
Protective clothing		

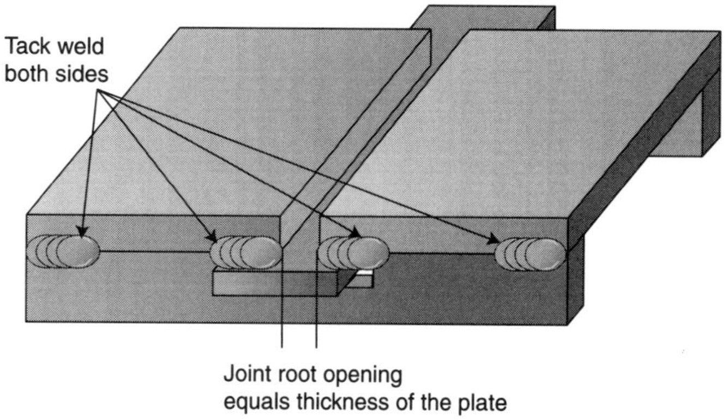

FIGURE 6-17 Preparation of square-groove weld with backing in flat position.

Welding Parameters

Amperes: 175 to 200

Voltage: 21 to 22

Electrode extension: ⅜ to ¾ inch

Directions

1. Clean the surfaces of the base metal with the grinder.
2. Make sure the power is off before plugging in the power source. Switch on power.
3. Adjust the gas flow from 35 cfh.
4. Set the voltage and set the wire speed; adjust by making test welds. Observe voltmeter and ammeter during welding to adjust welding parameters.
5. Clean the metal surfaces by grinding. Tack weld the two pieces together and set a joint root opening equal to the thickness of the plate. Tack weld the joint to the strong backs.
6. Watch the position of the gun to avoid having the weld pool run ahead. Use the backhand method to control this problem.
7. Remove all slag before laying the next pass.
8. Check with the instructor for evaluation, if necessary, before proceeding to the next exercise.
9. Pass the guided bend test before proceeding to the next exercise.

REVIEW QUESTIONS FOR UNIT 6

Multiple Choice

Choose the best answer to complete the statement.

1. Flat position:
   a. Deposits more metal.
   b. Is the position of most welding.
   c. Requires less time and energy to learn.
   d. Is a. and b. above.
   e. Is all of the above.
2. Necessary equipment for welding includes:
   a. Safety glasses.
   b. Welding gloves.
   c. Earplugs.
   d. All of the above.
3. Wire speed:
   a. Controls voltage.
   b. Controls amperage.
   c. Sets electrode speed.
   d. Is b. and c. above.
   e. Is all of the above.
4. Surfacing weld:
   a. Is another name for butt joint.
   b. Builds up surface of the plate.
   c. Is not a padding plate.

d. Is all of the above.
e. Is none of the above.

5. Ripples:
a. Are the patterns of each weld bead.
b. Are a type of slag.
c. If pointed, mean that travel speed is too fast.
d. Are a. and c. above.
e. Are none of the above.

6. Short circuiting transfer:
a. Produces slag.
b. Is a form of metal transfer.
c. Is used on thin metal.
d. Is b. and c. above.
e. Is all of the above.
f. Is none of the above.

7. Tack welds:
a. Are important in fabrication.
b. Are permanent weld beads.
c. Hold weldment in position.
d. Are a. and c. above.
e. Are all of the above.

8. Visual inspection:
a. Requires special technology.
b. Takes too much time.
c. Is the first test applied to any weld.
d. Is used only occasionally.
e. Is none of the above.

9. Intermittent welding:
a. Can be several tack welds.
b. Is a continuous weld bead.
c. Reduces heat input on thin metal.
d. Is a. and c. above.
e. Is a. and b. above.

10. Discontinuity is:
a. Porosity.
b. A slag inclusion.
c. A weld bead.
d. Both a. and b. above.
e. None of the above.

Short Answer

1. Briefly explain the functions of the American Welding Society.
2. Explain why most welding is completed in the flat position.
3. Explain when to use a flange.
4. What type of inspection is given to every weld?
5. Explain when a surfacing weld is used.
6. Explain the difference between the forehand and the backhand methods of welding.
7. Give two examples of welding parameters.
8. Describe the function of a tack weld.
9. Explain what can happen if all the slag is not removed before making another pass.
10. Define *downhill welding.*

ADDITIONAL ACTIVITIES

Riddles

1. I am a cap worn for safety reasons to protect the valve.
2. I am important because I comprise 78% of the air.
3. I am like a traffic cop, only I measure gas volume by flow.
4. When I am adjusted too low I can cause trouble for the contact tube.
5. If you were a beginning welder in my position, you would have to invest more time.
6. We set the standards for welding, and others must comply.
7. They threw away the mold when I was created by joining surfaces, both in flat and vertical planes.
8. The eyes have it when I am used to evaluate weld quality.
9. A troublemaker, I like to stop the flow of gas in the nozzle.
10. Without my support in holding things together, welds would be pulled every which way out of alignment.

Writing Skills

1. In a brief paragraph explain what to look for when making a visual inspection.
2. Make a list of the parts that comprise the gun, and describe the function of each part.

Math

A welder should know the common fractions in measuring. Most welders fabricate things within a tolerance of ⅛ inch, yet are able to measure within 1/16 inch when cutting and fitting for tack welding.

Reduce the following fractions to eighths and sixteenths.

1. 4/32 = ?/16 **2.** 8/64 = ?/16 **3.** 12/32 = ?/8
4. 16/64 = ?/8 **5.** 2/16 = ?/8 **6.** 6/16 = ?/8
7. 10/16 = ?/8 **8.** 7/8 = ?/16 **9.** 3/8 = ?/16
10. 8/16 = ?/8 **11.** 1/8 = ?/16

A welder should know the decimal equivalents for eighths and sixteenths. Please convert the following.

1/16 = 0.0625. example 1 ÷ 16 = 0.0625
1/8 = 0.125. example 1 ÷ 8 = 0.125

12. 3/16 = ? **13.** 5/16 = ? **14.** 15/16 = ? **15.** 7/16 = ?
16. 9/16 = ? **17.** 11/16 = ? **18.** 13/16 = ? **19.** 3/8 = ?
20. 5/8 = ? **21.** 6/8 = ? **22.** 7/8 = ? **23.** .250 = ?
24. .375 = ? **25.** .4375 = ? **26.** .045 = ?

UNIT 7

Joint Design

FIVE BASIC JOINTS

Choice of Joint Design

Welding begins with the simple idea of joining two pieces of metal together in a joint. There are five basic joints used in welding. Whenever two pieces of metal are joined, they fit into one of these basic joints. The five basic joints at the foundation of every welding project are the butt joint, tee joint, lap joint, corner joint, and **edge joint** (Figure 7-1).

Selection of the joint design is important in the fabrication of any project. The purpose is to achieve maximum strength in the joint with efficiency of the weldment. In other words, joints are designed to perform given tasks while requiring the least amount of welding. For example, the welder who is fabricating a container to hold a liquid would find the corner joint a more practical choice than a tee joint. A corner joint ensures a seal, saves material, yet looks attractive (Figure 7-2).

Some exercises in Unit 6 use the corner joint and the tee joint. These two joints plus the lap joint are the three easiest joints to weld. There should never be a mystery as to how much weld is enough. Size your welds equal to the thickness of the *thinnest* metal in the joint.

Fillet Weld

To become proficient at welding, every welder must learn about the anatomy of the particular welds being made. The fillet weld is a commonly used weld, as shown in a tee joint (Figure 7-3). The fillet weld joins the surfaces of two pieces, usually at right angles to each other. The fillet weld is popular because it does not require the added preparation necessary for the groove welds. Often the only preparation of the base metal for a fillet weld is grinding away surface materials that could interfere with the quality of the weld.

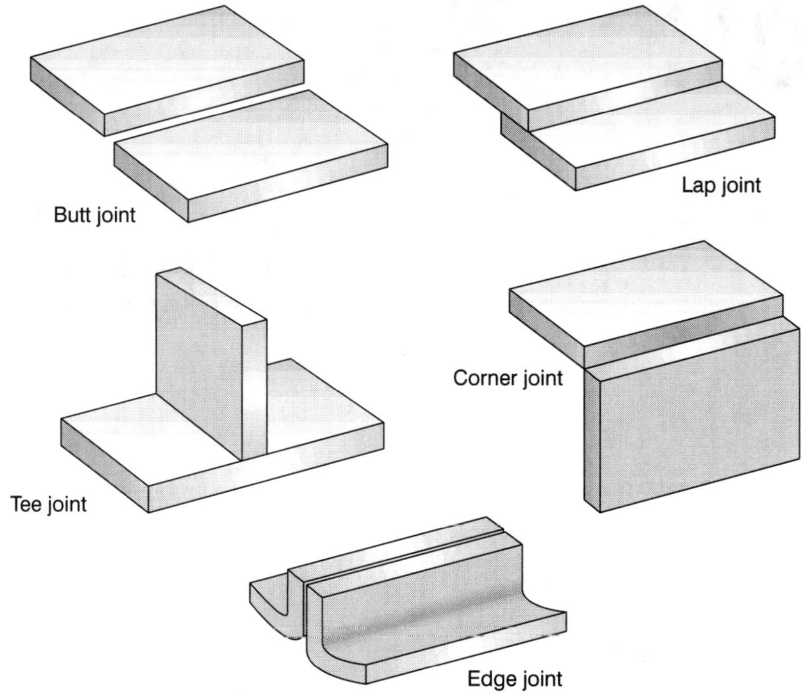

FIGURE 7-1 The five basic joints used in welding.

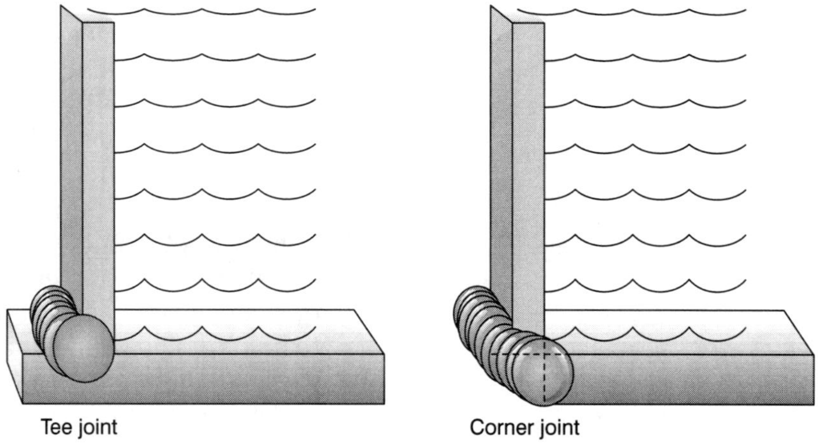

FIGURE 7-2 The use of the corner joint for sealed container applications is generally preferred over the tee joint.

Groove Welds

Used most often in the butt joint, the **groove weld** requires the most welding skill. The groove weld demands added preparation with the aim of achieving complete joint penetration. This means the weld can be viewed as penetrating completely (100%) to the back side of the joint. A groove weld is one of eight designs for the butt joint. This book will examine three of these designs—the **square-groove weld,** the **bevel-groove weld,** and the V-groove weld—with specifications from the AWS D1.1 Structural Welding Code-Steel.

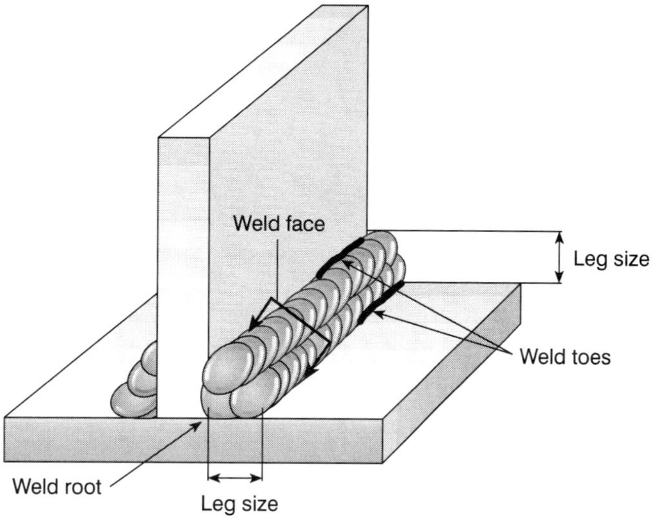

FIGURE 7-3 The parts of the fillet weld.

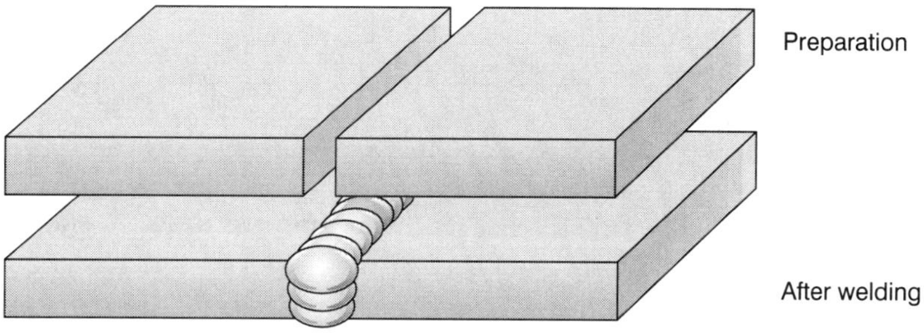

FIGURE 7-4 Square-groove weld.

The square-groove weld is for use in the butt joint without edge preparation. The square-groove weld is effective for joining sheet metal and plate with limited edge preparation. The corner joint and the butt joint are two applications of the square-groove weld. For GMAW and FCAW, the maximum recommended base metal thickness is ⅜ inch for complete joint penetration when making a square-groove weld from both sides of a butt joint (Figure 7-4). Other factors to be considered in welding plate with ¼-inch thickness are the choice of electrode, the welding position, amperage setting, and root opening.

The butt joint has to be prepared for welding a V-groove weld, or any of the other groove welds that require a bevel or the cutting away of some of the edge. The electric grinder, oxyacetylene torch, and mechanical beveler are three methods that can be used to bevel plate. The purpose of beveling plate is to aid root penetration into the joint (Figure 7-5). The **groove angle** is formed when two bevels form a butt joint. The **root opening** is the space between the pieces that form the joint. The **root face** can be the edge, not removed by beveling. Sometimes a root face is used to prevent melt-thru on the first pass, a situation in which too much of the base metal is melted away. The **groove face** is one of two

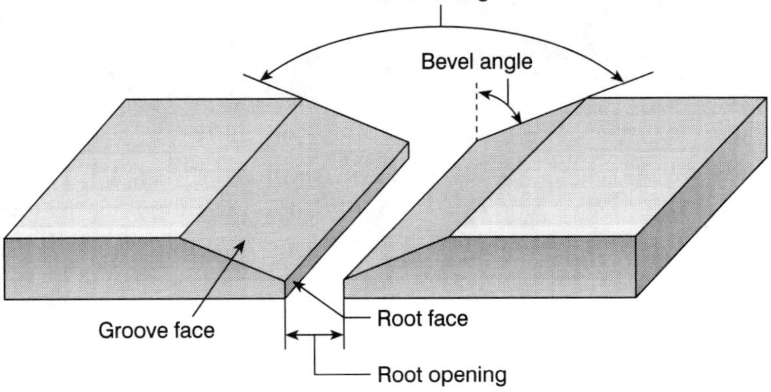

FIGURE 7-5 The parts of the V-groove butt joint.

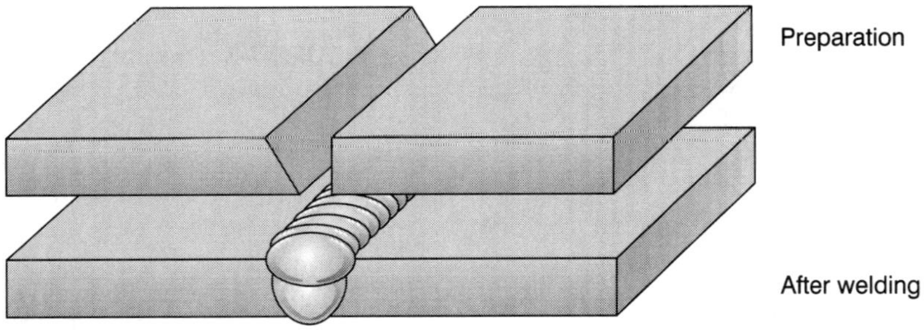

FIGURE 7-6 Bevel-groove weld.

surfaces included in the groove. These surfaces of the joint meet in a square-groove weld. In fact, the root face and the groove face are the same thing in the square-groove weld.

The bevel-groove weld is a weld in which the edge of one piece forming the joint is beveled. The bevel-groove weld is effective when preparing only one piece of the joint (Figure 7-6). For GMAW or FCAW, when welding from one side with backing material, either a 30° bevel angle with a root opening of 3⁄16 inch, a 45° bevel angle with a root opening of 1⁄4 inch, or a 30° bevel angle with a root opening of 3⁄8 inch is recommended.

The V-groove weld is a weld in which the edges of both pieces forming the joint are beveled (Figure 7-7). The V-groove weld is an effective weld when both pieces of the joint must be prepared. For GMAW or FCAW, the maximum recommended base metal thickness is unlimited for complete joint penetration. When welding from one side with backing material, the degree of the groove (two bevels) determines the size of the root opening: a 45° groove angle with a root opening of 1⁄4 inch, a 30° groove angle with a root opening of 3⁄8 inch, or a 60° groove angle with a root opening from 0 to 1⁄8 inch and a root face of 0 to 1⁄8 inch.

Many weldments fail because of welding on only one side of the joint. Observe how easy it is to break a weld by applying the force of a hammer on the side opposite the weld. One way to avoid this potential problem is to weld both sides of the joint or use groove welds with complete joint root penetration.

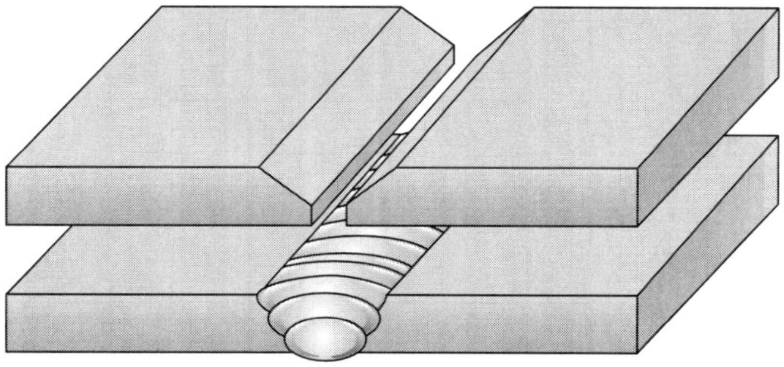

FIGURE 7-7 V-groove weld.

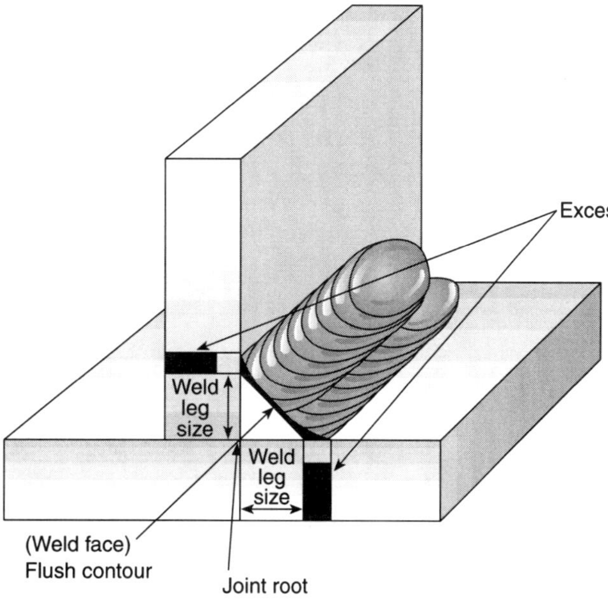

FIGURE 7-8 Measured size of a fillet weld is the leg size from the point of the flush contour.

Weld Size

The fillet **weld legs** help to establish the size of the fillet weld. Each fillet weld has two legs and two weld toes. The size of the fillet weld is determined by a diagonal line that connects each leg at the point of a flush contour, extending to the **joint root** (Figure 7-8). The joint root is the point where the surfaces of the base metal are closest. The contour of the weld face, whether **convex** or **concave,** is not a determining factor in the size of the weld. It does, however, establish the weld legs in the case of a concave fillet weld (Figure 7-9).

Other Welds

The American Welding Society lists 19 different welds used for joining metal. The fillet and groove welds receive most of the attention, but other welds are also important. The **plug weld** is

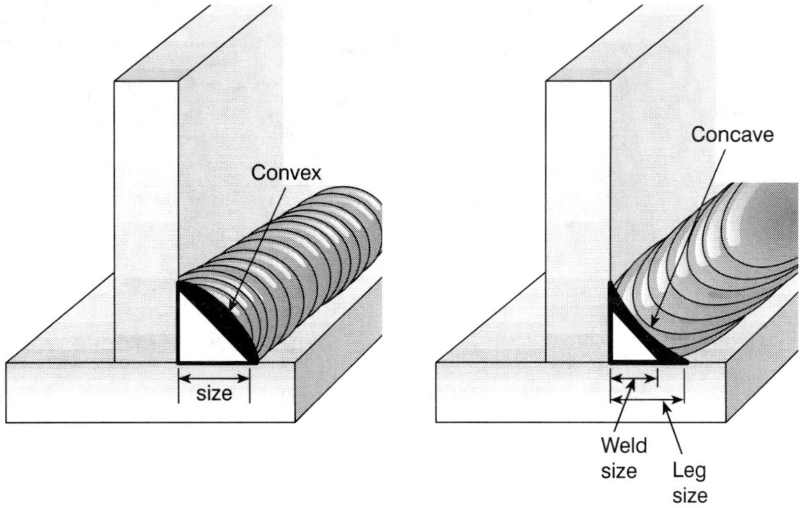

FIGURE 7-9 Profiles of the convex weld face and concave weld face.

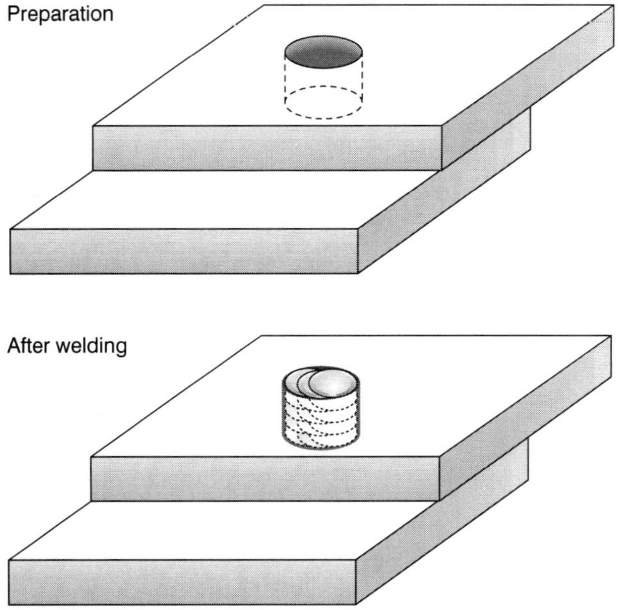

FIGURE 7-10 Plug weld in circular hole.

considered a remnant of the riveting age. The plug weld is a weld in a circular hole of one piece joined with a second piece in a lap joint (Figure 7-10). The size of the plug weld is determined by the diameter of the hole. The hole can be beveled for accessibility or increased holding strength, and is sometimes only partially filled with weld (Figure 7-11). The plug weld is used in applications when welding along the edge of a joint is not enough or is not practical.

The **slot weld** is another remnant of the riveting age. The slot weld is an elongated plug weld (Figure 7-12). The slot weld can also be beveled for accessibility and increased holding strength.

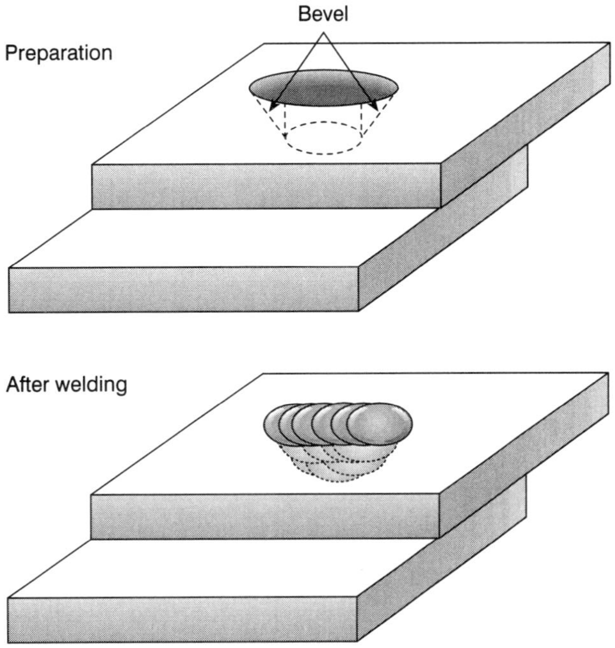

FIGURE 7-11 Plug weld in beveled hole.

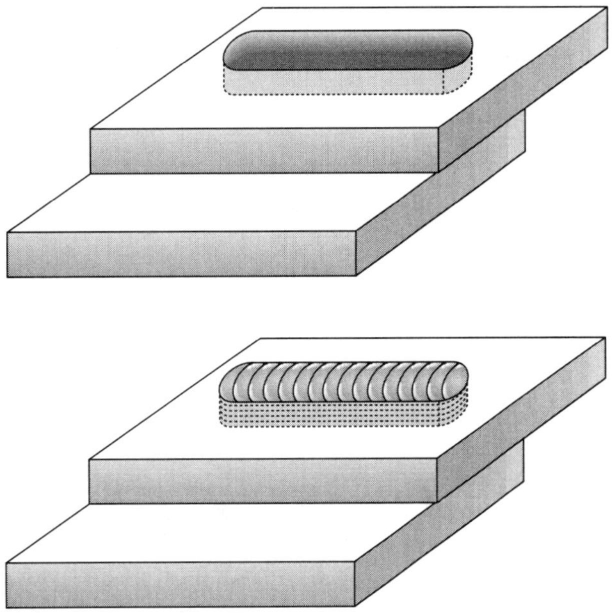

FIGURE 7-12 Slot weld in elongated hole.

Selecting a Joint Design

Selecting a joint design for welding depends on several factors. (1) The welder must determine the application of the joint and the forces that will act upon it. (2) The welder must choose joints that involve the least amount of welding and the least amount of preparation before welding. (3) The welder learns that thinner metals use smaller root openings, smaller bevel angles, and require less filler

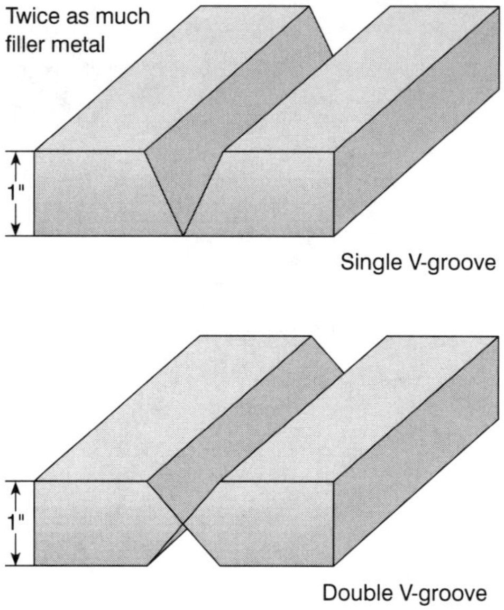

FIGURE 7-13 Double-groove weld versus single-groove weld; note that the single-groove weld will require more welding time.

> **ARC FLASH**
>
> A ¼-inch fillet weld 1 inch long using 70,000 series wire electrodes has a shear strength in the range of 2,600 pounds.

metal. (4) Finally, for thick plate, double-groove welds are preferred over single-groove welds (Figure 7-13). Double-groove welds require less filler metal, which means less time and less money.

The sizing of welds is also important. Remember that more welding requires more heat and more force pulling at the joint, and thus more distortion. Finding other ways to eliminate unnecessary welding can cut expenses and make a company more competitive.

Because overwelding causes undo stress on the joint, thus adding time and money, every effort should be made to size each weld to meet the requirements of the weldment. The maximum size weld should equal the thickness of the thinner member in the joint. The minimum size weld is determined by the application—those physical forces acting on the weldment. Always keep in mind that rules and common practices have their limitations. Physical forces can have extraordinary effects on a weldment. Consequently, limit welding to class projects in which weld failure will not cause any injuries.

> **SAFETY REMINDER**
>
> Weld failure can have devastating consequences, so avoid risk-taking when welding. At this stage of training, limit yourself to welding projects that have been engineered by others. Taking an advanced welding course can even be a rewarding experience that will benefit you the rest of your life.

REVIEW

1. Why is the fillet weld more popular than the groove weld?
2. What is the purpose for beveling a joint before welding?

3. Which joint, the bevel-groove weld or the V-groove weld, requires more preparation?
4. What are the results from overwelding?
5. How is the size of the fillet weld measured?

Lab Exercises

1. How many of the five joint designs can you locate in the shop?
2. How many of the following welds can you locate in the shop:
   a. fillet weld;
   b. square-groove weld;
   c. bevel-groove weld;
   d. V-groove weld
3. a. Locate fillet welds on structural beams around the shop.
   b. Try to determine their sizes.

REVIEW QUESTIONS FOR UNIT 7

Multiple Choice

Choose the *best* answer to complete the statement.

1. The five basic joints include:
   a. Square-groove.
   b. Edge.
   c. Lap.
   d. Butt.
   e. Answers a., b., c. above.
   f. Answers b., c., d. above.
2. ______ has (have) importance to joint design:
   a. Maximum strength with efficiency.
   b. As many different types of joints as possible.
   c. Maximum number of welds possible.
   d. Least amount of welding.
   e. Answers a. and c. above.
   f. Answers a. and d. above.
3. Parts of the fillet weld include:
   a. Weld toes and weld root.
   b. Weld legs and size.
   c. Weld face and weld length.
   d. Answers a. and b. above.
   e. Answers a. and c. above.
4. Parts of the butt joint include:
   a. Groove face and bevel angle.
   b. Root face and root opening.
   c. Groove angle and groove face.
   d. All of the above.
   e. Answers a. and b. above.
5. ______ is a method for beveling plate:
   a. Electric wrench.
   b. Oxyacetylene torch.
   c. Mechanical beveler.

d. Heavy-duty sander.
e. Answers a. and b. above.
f. Answers b. and c. above.

6. ______ requires the least amount of weld:
a. Square-groove weld.
b. Bevel-groove weld.
c. V-groove weld.
d. Answers a. and b. above.
e. None of the above.

7. The word *concave* is associated with:
a. Weld root.
b. Weld face.
c. Weld toes.
d. Joint root.
e. None of the above.

8. ______ is a remnant of riveting age:
a. Groove weld.
b. Plug weld.
c. Fillet weld.
d. Answers a. and b. above.
e. Answers b. and c. above.

9. Double-groove welds:
a. Are used on sheet steel.
b. Use more filler metal than single-groove welds.
c. Are used on thick plate.
d. Take more time and cost more money.
e. Include none of the above.

10. Overwelding:
a. Causes undo stress on the joint.
b. Costs more money.
c. Causes more force pulling on the joint.
d. Is a. and c. above.
e. Is a., b., c. above.

Short Answer

1. Name the five basic joints.
2. Name the parts of a fillet weld.
3. Explain when complete joint penetration is achieved.
4. Describe how the root face is different from the root opening.
5. What do fillet weld legs tell you?
6. Assuming that the fillet weld legs are the same, explain which requires more filler metal, a convex or concave weld face.
7. List three results of overwelding.
8. Decide which of the following welds requires more preparation (square-groove weld, bevel-groove weld, or V-groove weld), and explain why.
9. Explain why a corner joint design works better than a tee joint in fabricating a container to hold liquid.
10. Name the joint most often used with a V-groove weld.

ADDITIONAL ACTIVITIES

Matching

Connect each of the five basic joints with identifying characteristics.

Tee joint	a.	backing material
	b.	fillet weld
	c.	plug weld
Lap joint	d.	thin metal to limit distortion
	e.	preparation over ¼ inch for complete joint penetration
	f.	requires the most skill and preparation
Edge joint	g.	joining plate at 90° to each other
	h.	the joint disappears if the weld is removed by grinding
	i.	horizontal position most common for welding
Butt joint	j.	two pipes joined together on end
	k.	two pipes joined, telescoping out from each other
	l.	groove weld used most often
Corner joint	m.	**spot weld**

Writing Skills

1. In a brief paragraph explain what to consider when determining joint design.
2. Write a brief paragraph using the following terms: *weldment, concave, lap joint, bevel, fillet weld, fillet weld legs, ¼-inch thick,* and *weld face.*

Math

The job calls for fabricating a tank out of ⅛-inch sheet steel.
Remember that volume of a container = 5 width × 3 length × 3 height.

1. How many full sheets of 4′ × 8′ are needed to construct the tank, including one 3′ × 2½′ baffle (Figure 7-14)?

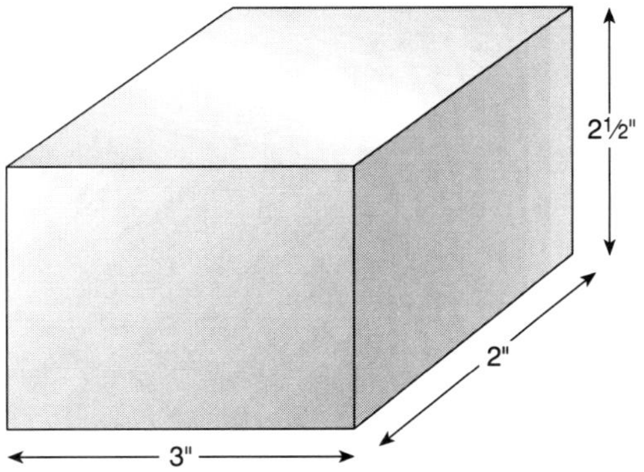

FIGURE 7-14 Tank to be constructed.

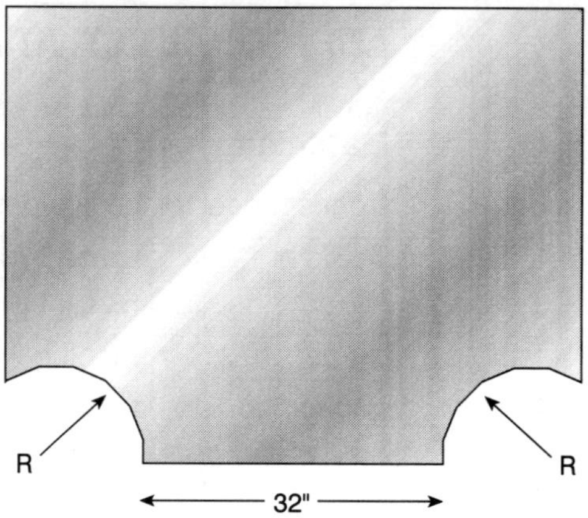

FIGURE 7-15 Baffle.

2. If a **press brake** is available to bend up sides to reduce welding and limit distortion, the total length of welding will be reduced from _____ to _____?
3. Determine the volume of the tank.
4. Convert your answer from question 3 to gallons. Remember that one gallon = 231 cubic inches.

   12″ · 12″ · 12″ = 1728 cubic inches or = 1 cubic foot
5. Determine the radius for the cutouts in the baffle (Figure 7-15).

Puzzlers

1. What determines going from a square-groove weld to a bevel-groove weld?
2. What are the considerations in deciding on a joint design?

UNIT 8

Electrode Wires and Shielding Gases

CHOICE OF ELECTRODES

Choosing an electrode actually begins with the selection of the welding process. In addition, the processes discussed in this book must find solutions for gases in the weld pool. When the amount of nitrogen in the weld pool exceeds limitations, the weld will develop a series of holes. The addition of special materials into the wire design can control the amounts of oxygen and nitrogen that enter the weld pool to improve weld quality. Each wire is manufactured to solve problems that arise in welding.

Is the welding job to be completed in an outside environment or in the controlled setting of a shop? FCAW that is self-shielded requires easy-to-use guns that are well suited for outside conditions where weather could affect the welding. GMAW that is gas shielding requires the protection from wind and drafts. Can the welding be completed by a single weld bead or will multiple weld beads be necessary? Wires have been designed for both applications. How thick is the base metal? Once again, the choice of the welding process and wire electrode will be affected by whether the base metal is thin or thick (Figure 8-1).

In choosing a welding process, the welder must consider what filler metals are available. The following provides some information to help in choosing between the different electrode wires.

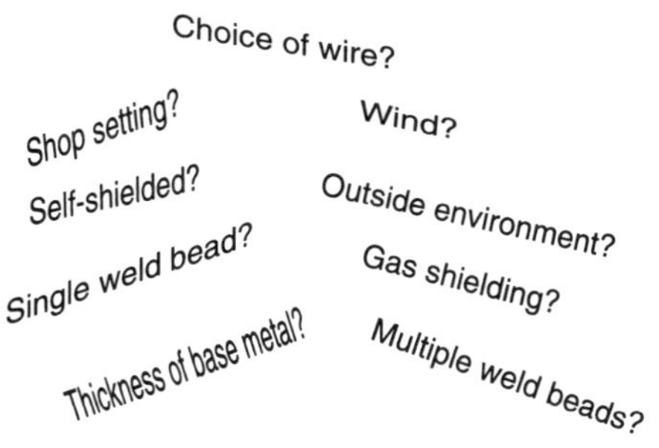

FIGURE 8-1 Welding involves decision making.

Types of Wire Electrodes

Three basic types of wire used in GMAW and FCAW are solid wires, flux cored wires, and metal cored wires. Solid wires are used with GMAW. The American Welding Society's specifications for mild steel filler metals are in the publication *ANSI/AWS/A5.18*.

ER70S-3

Regarding the "ER" prefix, the *E* designates electrode and the *R* designates a rod found in cut length. The next two digits, *70*, followed by the addition of three zeros (70,000), indicate the minimum tensile strength per square inch (psi). The *S* indicates a solid wire. The last number—which can be 2, 3, 4, 5, 6, or 7—provides information on the use and performance capabilities of the filler metal.

ER70S-2 is a highly deoxidized wire that can be used on carbon steels with thin coatings of rust or oxides. A **deoxidizer** combines with oxygen to form slag. This wirc can be used in all welding positions. Argon-oxygen, argon–carbon dioxide, and carbon dioxide are the recommended shielding gases.

ER70S-3 is a popular wire for both single and multiple weld beads, and should be used on clean base metals. It is preferred for welding galvanized base metals. Argon-oxygen and argon–carbon dioxide are recommended shielding gases.

ER70S-4 is a popular wire for structural steels, for example, A7 and A36. Argon-oxygen, argon–carbon dioxide, and carbon dioxide are the recommended shielding gases.

ER70S-5 is used on rusty steels; short circuiting transfer is not recommended. Argon-oxygen, argon–carbon dioxide, and carbon dioxide at high amperage settings are the shielding gases.

ER70S-6 is a general purpose wire effective on sheet metal. Deoxidizers make the act of welding base metals with thin coatings of rust and oxides practical. It is a general-purpose wire with a fluid weld pool that can be used in all welding positions. Argon-oxygen, argon–carbon dioxide, and carbon dioxide for short circuiting transfer are the recommended shielding gases.

ER70S-7 is a wire that can be used in several welding situations, including the weldment of heavy equipment and farm implements. Argon–carbon dioxide and carbon dioxide can be used in all welding positions.

The flux cored wires are used with FCAW. The American Welding Society's specifications for mild steel filler metals are in the publication *ANSI/AWS/A5.20.*

E71T-1M

Regarding the prefix, the *E* refers to *electrode.* The digit *7*, followed by the addition of four zeros (70,000), indicates the minimum tensile strength per square inch (psi). The next digit can be either a *0* or a *1*. A *0* indicates welding in the flat and horizontal positions. A *1* indicates welding in all positions. The *T* indicates a tubular wire with a flux inside the core. The second 1 (which can be 1 through 14), G, and GS provide information on the slag system, current, and shielding gas, respectively. The *M* indicates that the electrode has been labeled a 75 to 80% argon with the balance a carbon dioxide gas mixture. Some examples are as follows and refer to Table 8-1:

> E71T-1 is wire for single or multiple weld beads with direct current electrode positive (DCEP) in flat, horizontal, uphill, and overhead positions. This wire is designated for carbon dioxide and/or argon–carbon dioxide shielding gases; follow the manufacturer's recommendations. This wire is suitable for structural steel and earthmoving equipment.
>
> E71T-1M is a wire for single or multiple weld beads with DCEP in flat, horizontal, uphill, and overhead positions. This wire is designated for 75 to 80% argon with the balance carbon dioxide.
>
> E70T-1 is wire for single or multiple weld beads with DCEP in flat and horizontal positions. This wire is designated for carbon dioxide or argon–carbon dioxide shielding gas; follow the manufacturer's recommendations. The carbon steel must be cleaned before welding.
>
> E70T-2 is wire recommended for single weld bead applications. It contains deoxidizers for rust or mill scale and DCEP with carbon dioxide shielding gas in flat and horizontal positions.
>
> E70T-3 is a self-shielded wire recommended for DCEP, a single weld bead application, but is limited to less than ¼-inch thick. This electrode is for welding in flat and horizontal positions, and is successful on rapid welding of thin metal by automation.
>
> E70T-4 is a self-shielded wire recommended for DCEP multiple weld bead applications in flat and horizontal positions. It has a high deposition rate with low penetration for poor fit-ups.
>
> E70T-5 is a wire recommended for DCEP multiple weld bead applications in flat and horizontal positions. It uses globular transfer with carbon dioxide and/or argon–carbon dioxide shielding gases; follow the manufacturer's recommendations. The E70T-5 has low sensitivity to cracking.

The metal cored wires are used with GMAW. A new generation of metal cored wires has been developed to meet future welding needs. The metal cored wire contains a tubular wire with metal powder packed inside a sheath of thin carbon steel, also containing deoxidizers and arc stabilizers. These new metal cored wires have deposition rates from 10 to 30% greater than either solid wire or flux cored wires. They leave little to no slag so that cleaning between passes may not be necessary. They produce little spatter, up

TABLE 8-1 Useful information on last digit/AWS/A5.20 carbon and low-alloy steel electrodes for FCAW

AWS Classification	*Welding Current*	*Shielding Gas*	*Number of Passes*
EXXT-1	DCEP	Carbon dioxide	Multiple
EXXT-1M	DCEP	25-20% Carbon dioxide 75-80% argon	Multiple
EXXT-2	DCEP	Carbon dioxide	Single
EXXT-2M	DCEP	75-80% argon 25-20% Carbon dioxide	Single
EXXT-3	DCEP	None	Single
EXXT-4	DCEP	None	Multiple
EXXT-5	DCEP	Carbon dioxide	Multiple
EXXT-5M	DCEP	75-80% argon 20-25% Carbon dioxide	Multiple
EXXT-6	DCEP	None	Multiple
EXXT-7	DCEN	None	Multiple
EXXT-8	DCEN	None	Multiple
EXXT-9	DCEP	Carbon dioxide	Multiple
EXXT-9M	DCEP	75-80% argon 25-20% Carbon dioxide	Multiple
EXXT-10	DCEN	None	Single
EXXT-11	DCEN	None	Multiple
EXXT-12	DCEP	Carbon dioxide	Multiple
EXXT-12M	DCEP	75-80% argon 25-20% Carbon dioxide	Multiple
EXXT-13	DCEN	None	Single
EXXT-14	DCEN	None	Single
EXXT-G*	d*	d*	Multiple
EXXT-GS*	d*	d*	Single

G* is a new multiple weld bead electrode not presently covered under any classification.
GS* is a single weld bead electrode not presently covered under any classification.
d* properties as specified by the manufacturer.

to 95% less than other wires, and lower welding fume rates. They operate well with poor fit-ups and are less likely to cause melt-thru on thin metals. However, these metal cored wires are 20 to 25% more expensive than flux cored wires and the ultraviolet radiation is more intense, thus affecting the comfort of the welder.

FIGURE 8-2 Electrodes are available in various sizes and amounts.

The American Welding Society's specifications for mild steel metal core filler metals are in the publication *ANSI/AWS/A5.18.*

E70C-3M

The *E* designates electrode. The next two digits, *70*, followed by the addition of three zeros (70,000), indicate the minimum tensile strength per square inch (psi). The *C* indicates a composite metal cored wire. The *3* refers to impact strength equal to 20 foot-pounds at 0°F. A *6* indicates 20 foot-pounds at –20°F. The last letter refers to the shielding gas; *M* means 75 to 92% argon with the balance in carbon dioxide, whereas a *C* means 100% carbon dioxide. Manufacturers have developed many brand names that fit under the following two examples.

E70C-3C is a metal cored wire designed to replace solid, mild steel wires where the user wishes to employ lower-cost carbon dioxide. Lower spatter rates with DCEN are recommended.

E70C-6M is a metal cored wire designed to make multiple passes without cleaning in between. It requires 75 to 92% argon with 8 to 25% carbon dioxide.

Electrode Sizes

The electrodes or wires are designed to be consumed at high speeds, which occurs when the wire is fed off 10- to 250-pound spools. The wire can also unwind off drums and reels as large as 750 to 1,000 pounds, or off small, 1- to 2-pound spools used in the more portable spool guns (Figure 8-2).

The diameter of the electrode can vary with smaller wire sizes of 0.035 and 0.045 popular with GMAW. While diameter can be as large as ⅛ inch, FCAW uses popular wire sizes of 0.045 inch, 1⁄16 inch, and 3⁄32 inch (Figure 8-3).

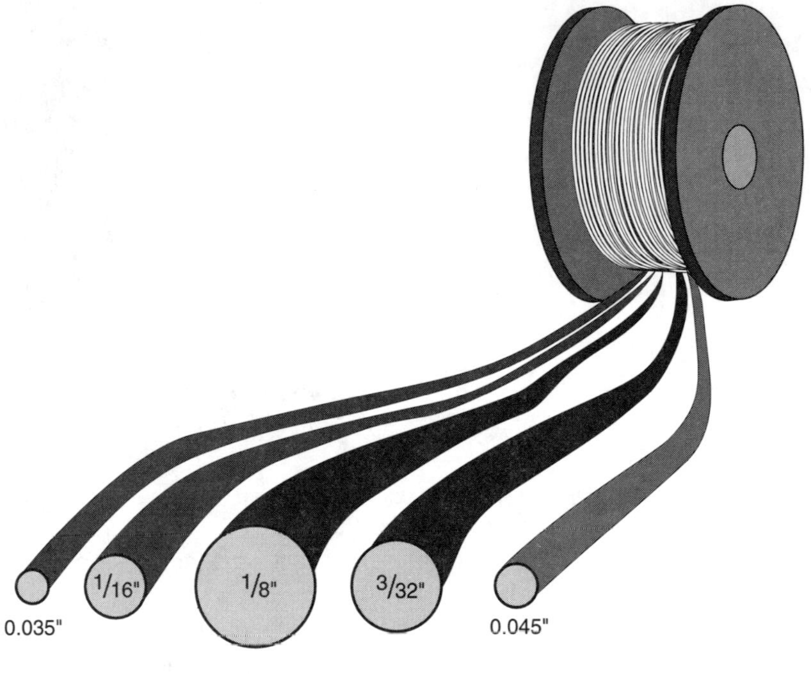

FIGURE 8-3 Some popular electrode diameters.

ARC FLASH

Always keep your head out of the smoke stream.

SAFETY REMINDER

A new generation of lightweight, fume-extracting welding guns is now available. Occupational Safety and Health Administration (OSHA) regulation concerning the environment for welding limits the iron oxides in the breathing zone to 10 milligrams per cubic meter (150 grains per 1.307 cubic yards).

Storage of Electrode Wires

Flux cored arc welding electrodes must be protected from their tendency to pick up moisture in the atmosphere, whether it be rain, snow, spray, or humidity. Only hermetically sealed containers are completely safe. Reconditioning is not recommended; always follow the manufacturer's instructions. Remember that electrode wire is packaged to prevent contamination, so keep the wire in its original packaging until needed. Because these electrodes are designed to be used in a clean and dry environment, try to protect the wire when ideal conditions are not possible.

Variety of Electrode Wires

The basic welding instruction exercises in this book are designed for mild steel base metals and carbon steel electrodes; however, other metals can also be used. See tables 8-2 and 8-3. Information on electrode selection can be found in publications giving specifications developed by the American Welding Society (AWS), 550 N.W. LeJeune Road, Miami, Florida 33126.

TABLE 8-2 GMAW electrodes

Type of Base Metal	*AWS Specification*	*AWS Electrode Specification*
Carbon steel	A5.18	ER70S-2, ER70S-3, ER70S-4, ER70S-5, ER70S-6, ER70S-7
Low-alloy steel	A5.28	ER80S-D2, ER80S-G, ER80S-Ni1, ER90S-G, ER100S-G, ER100S-1, ER110S-G, ER120S-1
Aluminum alloys	A5.10	ER4043, ER5356, ER5554, ER5556, ER5183
Copper alloys	A5.7	ERCu, ERCuSi-A, ERCuSn-A, ERCuNi, ERCuAl-A2
Magnesium	A5.19	ERAZ61A, ERAZ92A
Nickel alloys	A5.14	ERNi, ERNiCu-7, ERNiCrFe-5
Stainless steel	A5.9	ER308, ER308L, ER308L-Si, ER309, ER309L, ER310, ER312, ER316, ER316L, ER316L-Si, ER317L, ER320, ER320LR, ER347
Titanium	A5.16	ERTi-1, ERTi-2, ERTi-3, ERTi-4, ERTi-6Al-4V, ERTi-0.2Pd

TABLE 8-3 FCAW electrodes

Type of Base Metal	*AWS Specification*	*AWS Electrode Specification*
Carbon steel	A5.20 self-shielding	E70T-4, E70T-7, E70T-8, E70T-10, E71T11, E71T8-K6, E71T8-Ni1, E71T8-Ni2, E71T-11, E71T-GS
Carbon steel	A5.20 gas shielding	E70T-1, E70T-2, E70T-5, E71T-1
Carbon steel	A5.18 metal cored	E70C-3C, E70C-6M
Low-alloy steel	A5.29 self-shielding	E91T8-G, EG82T-G
Low-alloy steel	A5.29 gas shielding	E80T1-K2, E80T1-W, E81T1-B2, E81T1-Ni1, E81T1-Ni2, E91T1-K2, E110T1-G, E110T5-Kr
Low-alloy steel	A5.29 metal cored	E80C-G, E110C-G, E120C-G

Lab Exercises

1. Explain the AWS classification for two electrodes used in your shop.
2. If your shop has solid wires, flux cored wires, and/or metal cored wires, briefly describe their differences.
3. If AWS specification publications are available, in a paragraph summarize the information provided.

CHOICE OF SHIELDING GASES

Over 50 gas mixtures are available for welding. A welder can therefore become confused by the sales and marketing involved in selecting a shielding gas. Remember that a primary task of any shielding gas is to protect the weld pool as it cools from contaminants in the air. Air consists of 21% oxygen and 78% nitrogen with the final 1% trace amounts of several other gases. Other functions of the shielding gas include:

1. to stabilize the arc;
2. to establish the range of voltage;
3. to determine metal transfer;
4. to affect the amount of penetration; and
5. to assist in the appearance of the weld bead.

The two types of gases are inert or inactive and active. An inert or inactive gas such as argon is stable and does not react with oxygen or form any other compounds readily. The active gases such as carbon dioxide and oxygen react with other elements to form compounds. Argon, carbon dioxide, and oxygen are the three primary shielding gases used in different combinations.

Argon in combination with carbon dioxide is preferred for welding carbon steels, although 100% carbon dioxide and 75% argon to 25% carbon dioxide are popular choices for short circuiting transfer (GMAW-S). The use of carbon dioxide, although less expensive, requires a higher voltage for GMAW-S that makes for a more unstable arc and increased spatter with greater possibility for melt-thru on thin metal with more fumes. An argon–carbon dioxide mix provides a broad range of parameters with better arc stability, and is highly successful on thin gauge metal from 20 gauge to 11 gauge.

Tables 8-4 to 8-11 provide some suggested parameters in setting a power source for the various electrode wires used in the different welding processes. When setting the power source properly, lay some practice weld beads.

TABLE 8-4 Setting the parameters (approximate) for short circuiting transfer (GMAW-S) 0.030 diameter wire

Steel Thickness	*Shielding Gas*	*Gas Setting Cubic ft/hr*	*Amperes*	*Voltage*
24 gauge	CO_2	15 to 20 cfh	30 to 40	15 to 16
22 gauge	CO_2	15 to 20 cfh	40 to 50	16 to 17
20 gauge	CO_2	15 to 20 cfh	50 to 60	17 to 18
18 gauge	CO_2	15 to 20 cfh	70 to 80	18 to 19
16 gauge	CO_2	15 to 20 cfh	90 to 110	19 to 20

TABLE 8-5 Setting the parameters (approximate) for short circuiting transfer (GMAW-S) 0.030 diameter wire

Steel Thickness	*Shielding Gas*	*Gas Setting Cubic ft/hr*	*Amperes*	*Voltage*
24 gauge	75% Ar/25% CO_2	15 to 20 cfh	30 to 40	15 to 16
22 gauge	75% Ar/25% CO_2	15 to 20 cfh	40 to 50	15 to 16
20 gauge	75% Ar/25% CO_2	15 to 20 cfh	50 to 60	15 to 16
18 gauge	75% Ar/25% CO_2	15 to 20 cfh	70 to 80	16 to 17
16 gauge	75% Ar/25% CO_2	20 to 25 cfh	90 to 110	17 to 18

TABLE 8-6 Setting the parameters (approximate) for short circuiting transfer (GMAW-S) 0.035 diameter wire

Steel Thickness	*Shielding Gas*	*Gas setting Cubic ft/hr*	*Amperes*	*Voltage*
1/16 inch	CO_2	20 to 25 cfh	80 to 110	18 to 20
1/8 inch	CO_2	20 to 25 cfh	120 to 150	19 to 20
1/4 inch	CO_2	20 to 25 cfh	140 to 160	19 to 21
5/16 inch	CO_2	20 to 25 cfh	200 to 210	21 to 22
3/8 inch	CO_2	20 to 25 cfh	220 to 250	22 to 24

TABLE 8-7 Setting the parameters (approximate) for short circuiting transfer (GMAW-S) 0.035 diameter wire

Steel Thickness	*Shielding Gas*	*Gas setting Cubic ft/hr*	*Amperes*	*Voltage*
1/16 inch	75% Ar/25% CO_2	20 to 25 cfh	95 to 100	18 to 20
1/8 inch	75% Ar/25% CO_2	20 to 25 cfh	140 to 160	19 to 21
1/4 inch	75% Ar/25% CO_2	20 to 25 cfh	150 to 180	19 to 21
5/16 inch	75% Ar/25% CO_2	20 to 25 cfh	170 to 210	21 to 22
3/8 inch	75% Ar/25% CO_2	20 to 25 cfh	220 to 250	22 to 24

TABLE 8-8 Setting the parameters (approximate) for spray-arc transfer (GMAW-SP) 0.045 diameter wire

Steel Thickness	*Shielding Gas*	*Gas Setting Cubic ft/hr*	*Amperes*	*Voltage*
⅛ inch	98% Ar/2% O_2	40 to 50 cfh	160 to 170	13 to 24
3/16 inch	98% Ar/2% O_2	40 to 50 cfh	180 to 190	24 to 25
¼ inch	98% Ar/2% O_2	40 to 50 cfh	200 to 250	24 to 26
⅜ inch	98% Ar/2% O_2	40 to 50 cfh	300 and up	26 to 27
½ inch	98% Ar/2% O_2	40 to 50 cfh	310 and up	29 to 30

TABLE 8-9 Setting the parameters (approximate) for flux cored arc welding, self-shielded

Electrode Diameter	*Electrode Extension*	*Amperes*	*Voltage*
.045 inch	⅛ to 1 inch	60 to 400	13 to 34
1/16 inch	¼ to 1 inch	110 to 500	14 to 39
3/32 inch	½ to 1¼ inch	200 to 600	17 to 39

TABLE 8-10 Setting up for flux cored arc welding, gas shielding (shielding gas: carbon dioxide and argon–carbon dioxide mixtures to 75% argon)

Electrode Diameter	*Electrode Extension*	*Amperes*	*Voltage*
.045 inch	¼ to 1 inch	150 to 400	18 to 34
1/16 inch	½ to 1 inch	170 to 500	22 to 39
3/32 inch	¾ to 1½ inch	300 to 600	26 to 40

TABLE 8-11 Setting the parameters for flux cored arc welding, metal cored wires (shielding gas: 75 to 92% argon/25 to 80% carbon dioxide)

Electrode Diameter	*Electrode Extension*	*Amperes*	*Voltage*
.045 inch	¼ to ¾ inch	75 to 350	15 to 35
1/16 inch	⅜ to 1 inch	150 to 500	17 to 35
3/32 inch	¾ to 1¼ inch	350 to 580	26 to 37

SAFETY REMINDER

Remember to use care when transporting any high-pressure gas cylinder. Always be sure the cylinder cap is positioned correctly before moving to avoid accidental damage of the valve, which can turn a cylinder into a dangerous projectile.

Lab Exercises

1. What are the shielding gas mixtures used in your shop?

2. List the characteristics of each gas to explain their use.

REVIEW QUESTIONS FOR UNIT 8

Multiple Choice

Choose the *best* answer to complete the statement.

1. The ______ is involved in choosing an electrode.
   a. Environment.
   b. Welding process.
   c. Number of passes.
   d. Size of wire spool.
   e. All of the above.
   f. Answers a., b., and c. above.

2. ______ are a type of electrode.
   a. Solid wires.
   b. Metal cored wires.
   c. Deoxidized wires.
   d. Flux cored wires.
   e. All of the above.
   f. Answers a., b., and d. above.

3. When interpreting ER70S-5:
   a. *70* means 70,000 tensile strength.
   b. *S* means self-shielded.
   c. *5* means information on performance.
   d. All of the above.
   e. Answers a. and b. above.
   f. Answers a. and c. above.

4. When interpreting E70T-7:
   a. First *7* means 70,000 tensile strength.
   b. *0* means flat/horizontal positions.
   c. *T* means tubular.
   d. Second *7* means carbon dioxide.
   e. All of the above.
   f. Answers a., b., and c. above.

5. Metal cored wires:
   a. Are tubular.
   b. Have greater deposition rate than solid wires.
   c. Produce little slag.
   d. Are all of the above.
   e. Are none of the above.

6. When interpreting E70C-3M:
   a. *70* means 70,000 tensile strength.
   b. *C* means carbon dioxide.
   c. *3* means welding positions.
   d. *M* means metal cored.
   e. All of the above.

7. Diameter of electrode includes:
   a. 0.45.
   b. ⅛ inch.

c. 3/32 inch.
d. All of the above.
e. Answers b. and c. above.

8. Shielding gas is:
a. Oxygen alone.
b. Argon.
c. Nitrogen.
d. Air.
e. Answers a. and b. above.

9. Carbon dioxide:
a. Is a shielding gas.
b. Is the most expensive of shielding gases.
c. Causes more spatter.
d. Is all of the above.
e. Is a. and c. above.

Short Answer

1. Describe the difference between an active and an inert gas.

2. Describe a major difference to be considered between self-shielded FCAW and GMAW outside.

3. Name a purpose for using a deoxidizer in wire electrodes.

4. Tell why it is important to know the type of current (DCEN or DCEP) being used.

5. List some advantages of metal cored wires.

6. Explain the concern of OSHA as mentioned in this unit.

7. Describe how to limit contamination of the wire electrode.

8. List four other base metals besides carbon and low-alloy steel that can be welded.

9. Describe the primary task of any shielding gas.

ADDITIONAL ACTIVITIES

Crossword Puzzle

Across

1. Wire not designed for a single pass.

2. Developed specifications for filler metals.

3. Type of steel.

4. Type of wire with flux inside core.

5. Metal core wires reduce this unhealthy by-product.

6. Element (78% of the air) that must be kept out of the weld pool.

7. Stable gas.

8. Created during welding.

Down

1. Type of wire.

2. Shielding does not require a gas.

3. Shielding gas.

4. Welding position.

5. One of the five basic joints.

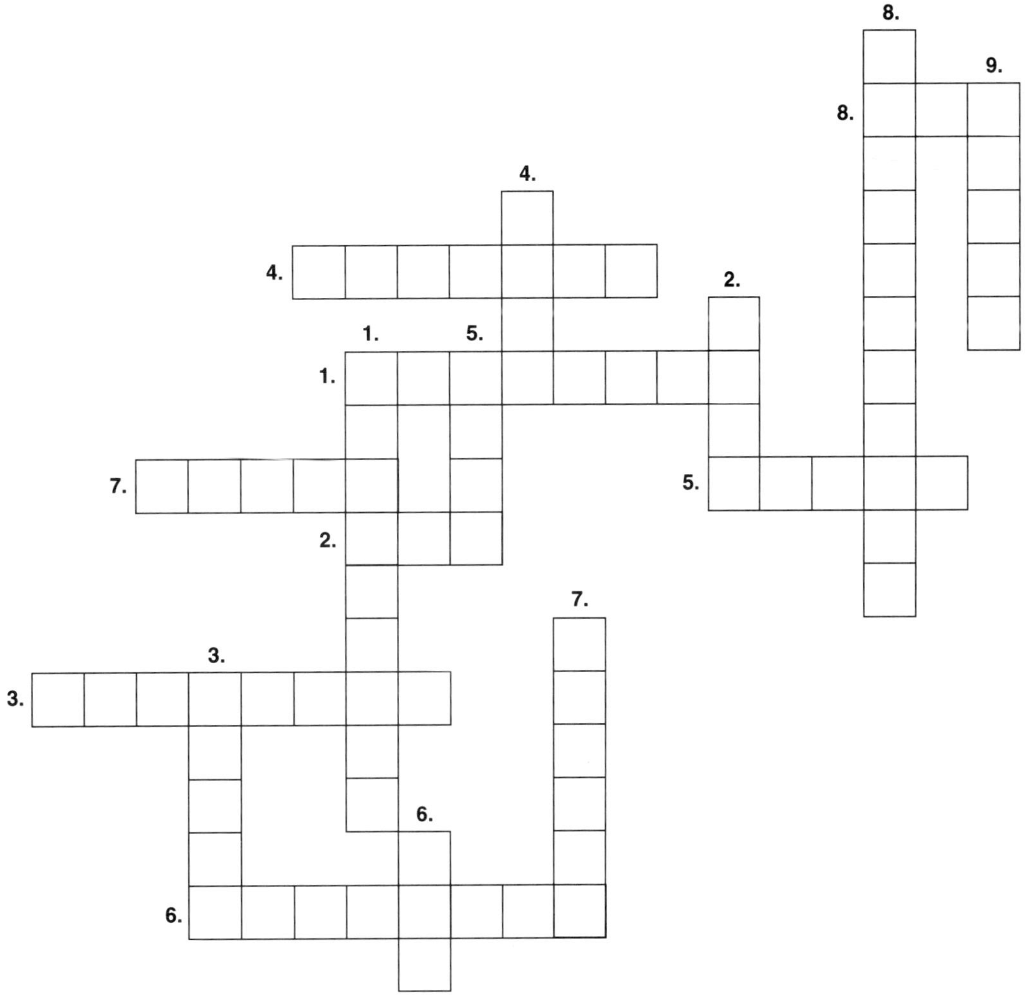

6. Carbon dioxide abbreviated.
7. Example of active gas.
8. Used to set power source for welding.
9. Measurement of a gas by volume.

Writing Skills

1. In a paragraph, describe the purpose of a shielding gas.
2. In a paragraph, describe the differences between a solid wire, a tubular wire, and a metal cored wire.
3. In a paragraph, explain the method in choosing one electrode over another.

Math

Volume is a cubic measurement.

12″ = 1 foot	12″ × 12″ = 144 square inches
12″ × 12″ × 12″ = 1,728 cubic inches	1′ (foot) × 1′ = 1 square foot

1′ × 1′ × 1′ = 1 cubic foot   1 cubic foot = 1,728 cubic inches

Surface area measurements = L (length) × W (width)

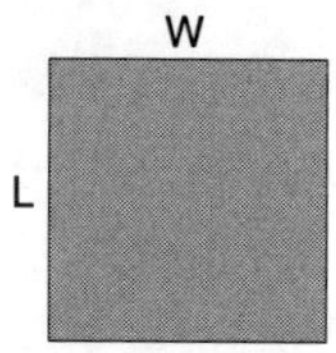

1. Find the area of illustration.
   L = 14 feet; W= 8 feet.
2. Find the area.
   L = 10 feet; W = 8 feet 9 inches.
3. How are the answers for the above questions expressed in area?
4. Find the volume of illustration.
   L = 8 feet; W = 6 feet; H (height) = 7 feet
5. Find the volume.
   L = 5 feet; W = 4½ feet; H = 6 feet

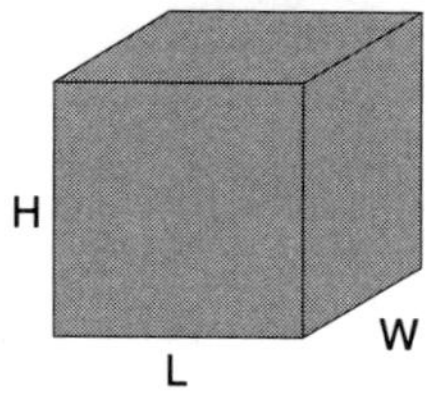

The weight of gas is a measurement of its density, and is expressed as weight in pounds per cubic foot. Gas density × volume (cubic feet) = weight of the gas.

Gas	*Gas Density per Cubic Foot*
argon	0.103
carbon dioxide	0.114
oxygen	0.083
helium	0.010
hydrogen	0.005

6. What is the gas weight if the cylinder contains 244 cubic feet of oxygen?
7. What is the gas weight if the cylinder contains 213 cubic feet of helium?
8. What is the gas weight if the cylinder contains 104 cubic feet of hydrogen?
9. What is the gas weight if the cylinder contains 242 cubic feet of argon?
10. What is the gas weight if the cylinder contains 213 cubic feet of carbon dioxide?
11. What is the gas weight if the cylinder contains 150 cubic feet of argon?
12. What is the gas weight if the cylinder contains 150 cubic feet of oxygen?

Puzzlers

1. What are the factors in choosing GMAW over FCAW?
2. What are the factors in choosing FCAW over GMAW?

UNIT 9

Heat and Distortion

EFFECTS OF WELDING

As stated in Unit 3, steels are selected for their properties. Medium-carbon steels are used in applications requiring moderate hardness and wear resistance. Low-alloy steels are used in applications subject to a higher degree of toughness and hardness with greater tensile strength than low-carbon steels. Medium-carbon and low-alloy steels are found in industries such as mining, agriculture, and shipbuilding.

Both medium-carbon and low-alloy steels require taking precautions when welding. The concern is that **martensite** can form in the heat-affected zone (HAZ), which can lead to cracking and make preheat and/or postheat necessary. Martensite is a hard, pin-like grain structure that forms when some steels cool too rapidly. Martensite forms in steel when the cooling rate of a welded joint is so fast that a more ductile **grain structure** is unable to form.

When preheating is necessary, the preheat temperature of the base metal for welding is determined by such factors as the type of steel. This temperature can range from 150° to 550°F; such a temperature is practical without requiring specialized equipment. Preheating helps to slow the rate of cooling to prevent the formation of martensite.

Expansion and Contraction

All metals are affected by heat differently. One difference is in the way metals expand when heated. This expansion, which varies with the type of metal (aluminum versus steel, for example), can be measured and is called the **coefficient of linear expansion.** Note in Table 9-1 that steel expands at approximately one-half the rate of aluminum. This means that aluminum is nearly twice as

TABLE 9-1 Expansion coefficient for common metals per rise in 1° of temperature

Metal	*Expansion Coefficient per Inch*
Aluminum	0.0000123
Copper	0.0000088
Steel	0.0000063
Cast iron	0.0000056
Stainless steel	0.000009

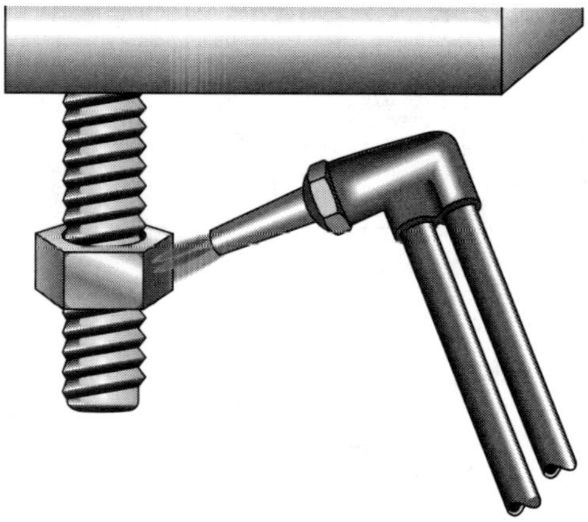
FIGURE 9-1 Heat expands metal, making removal of a nut easier.

sensitive to heat and more readily subject to distortion than steel. A person can observe the effects of this expansion when hearing a metal barrel pop on a hot, sunny day or a kitchen stove make a noise as it cools. Knowledge of how heat causes expansion can be used to the welder's advantage. One application of controlled heating is to apply the flame of a torch to a nut on a rusted bolt. The heat will cause the nut to expand so it can be twisted off without difficulty (Figure 9-1). Having this type of knowledge makes it easier on the tools and on a person's back. Obviously on large welding projects, measurements will change depending on the temperature at a given time of day.

Whereas metal expands when heated, it also contracts on cooling. If metal is heated evenly and allowed to cool evenly, it will return to its original shape if not restricted. For example, put a bar 1 inch diameter by 6 inches long into a vice and heat evenly to orange-red, then let the bar cool. The bar having been restricted should fall from the vice upon cooling, and be thicker but shorter than before (Figure 9-2). This bar becomes thicker due to distortion. In welding, metal never returns to its original shape because heating is always uneven as shown in Figure 9-2. A welder quickly learns how heat can distort metal by attempting to weld together a frame without placing tack welds (Figure 9-3). A frame may move out of position so distorted that it becomes impossible to bring back together without cutting out the welds.

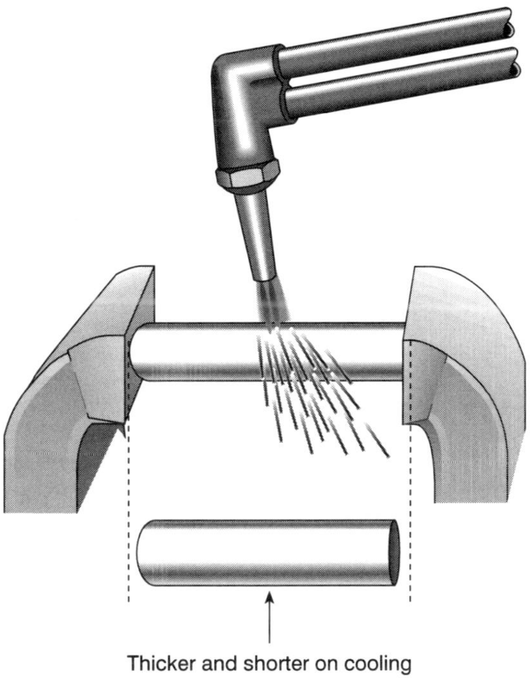

FIGURE 9-2 Restricted heating can affect the size of an object on cooling.

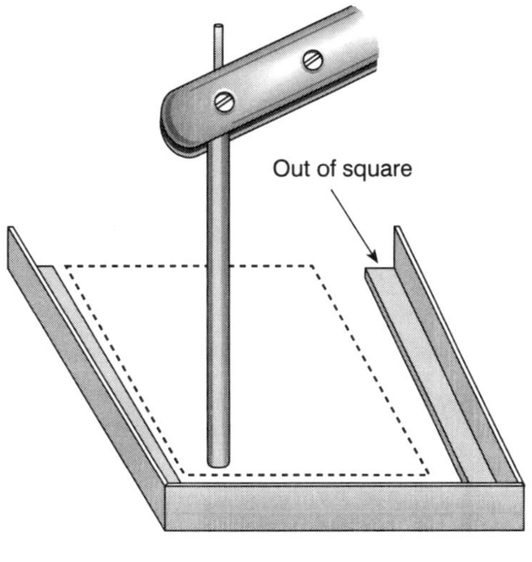

FIGURE 9-3 The effects of distortion caused by failure to use tack welds.

Distortion and the Amount of Welding

The greater the amount of heat generated in the joint, the greater the possibility of distortion. Any time the welder can limit the size of a weld, less heat will result in less distortion. Compare a ½-inch fillet weld with a 1-inch fillet weld (Figure 9-4). A 1-inch fillet weld has twice the strength of a ½-inch fillet weld, but requires four times the filler metal. A welder must consider the added stress that extra filler metal places on the joint. In most applications the thickness of the base metal will determine the size of the weld.

Welding produces stress in metal—even metal that has not been welded is stressed through the manufacturing process. Observe this stress by using a hacksaw to cut through a small length of tubing or pipe. Before the cut is complete, the blade will

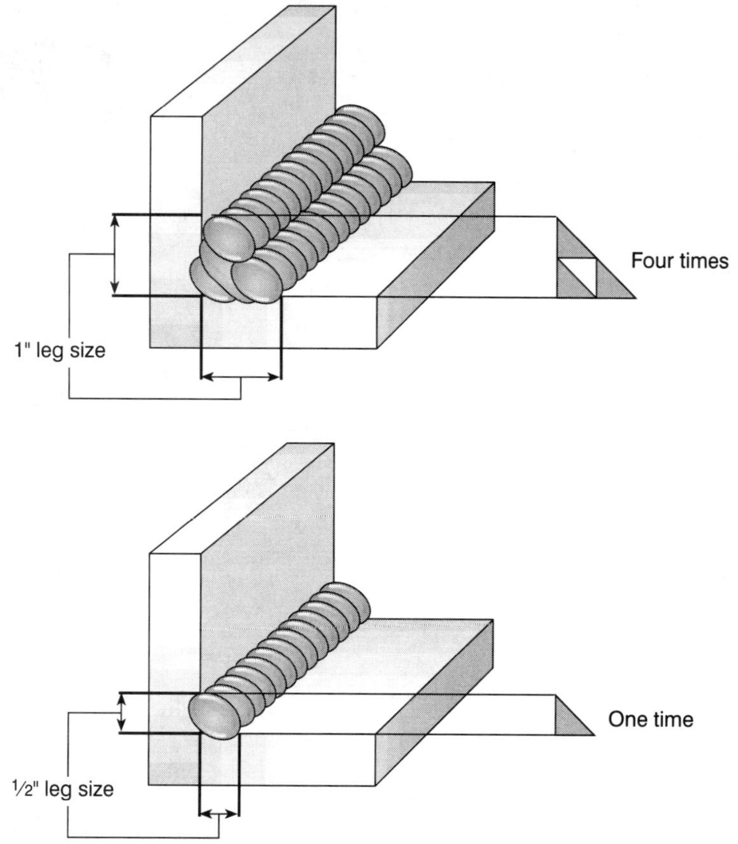

FIGURE 9-4 One-inch weld compared with ½-inch weld requires four times more filler metal.

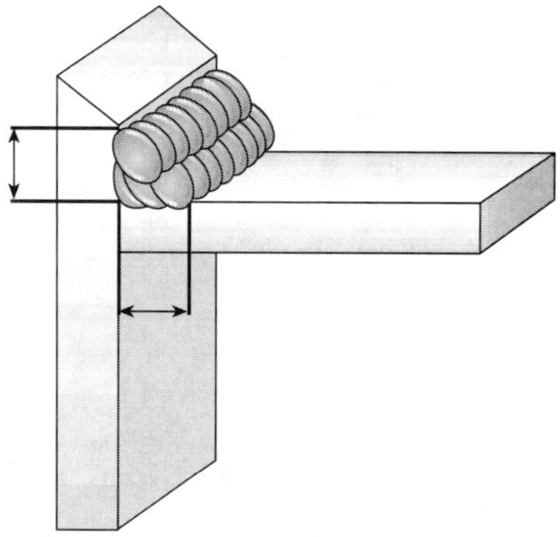

FIGURE 9-5 Equal leg fillet weld.

become pinched as the cut closes in an effort to relieve some of the stress built into the tubing or pipe during its manufacture. One way to limit stress is to do no more welding than necessary.

Equal leg fillet welds should be used in most applications (Figure 9-5). When the legs are of unequal size, the strength of the weld is decided by the smaller of the two legs. Notice that doubling the size of one leg (½-inch leg by 1-inch leg) will double the filler metal required, but add only one-tenth to the **effective throat** (Figure 9-6). The effective throat is measured as the shortest distance from the **weld root** to the weld face at the point of a **flush contour.**

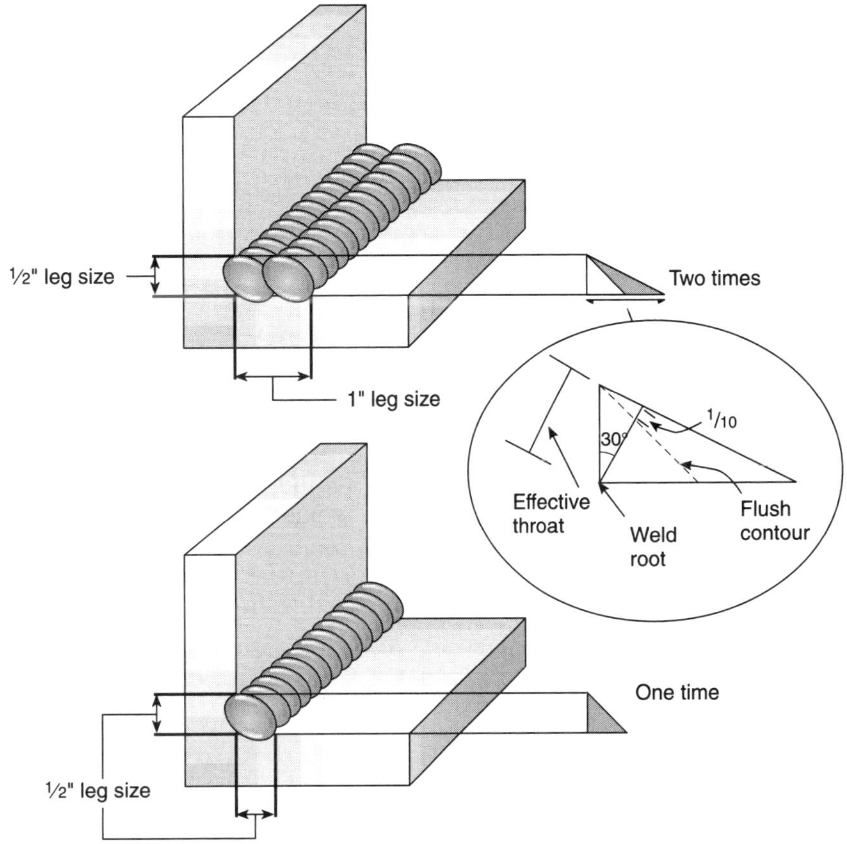

FIGURE 9-6 Unequal-legged ½-inch by 1-inch fillet weld compared with equal-legged ½-1-inch fillet weld requires twice the filler.

> **ARC FLASH**
>
> Although many common stainless steels are nonmagnetic, 400 series stainless steels are magnetic.

Setup for Welding

A welder must learn how to anticipate the effects welding will have on a given weldment and must be prepared to handle any distortion. Upon cooling, the molten weld pool will contract, pulling on the joint. Keep the following points in mind when welding several pieces together on any fabrication project.

1. Tack weld all pieces together. Use clamps to hold the weldment in position to keep the tack welds from pulling (Figure 9-7).
2. Try to spread out the welding to limit the concentration of heat at any one joint. If the joint is allowed to cool, the effects of heat can be reduced.
3. Do less welding and as quickly as possible. Use as large an electrode as possible to reduce the number of passes.
4. Keep the filler metal to a minimum. Do not overweld.
5. Use a faster travel speed during welding to reduce heat input; however, it can cause rapid cooling.
6. Preheating when necessary, will provide a more even heat to slow cooling so as to reduce the shrinkage in the welds (Figure 9-8).
7. Cooling between layers in many-layered welds can reduce distortion; however, this can change the grain structure of steels that are sensitive to the formation of martensite.

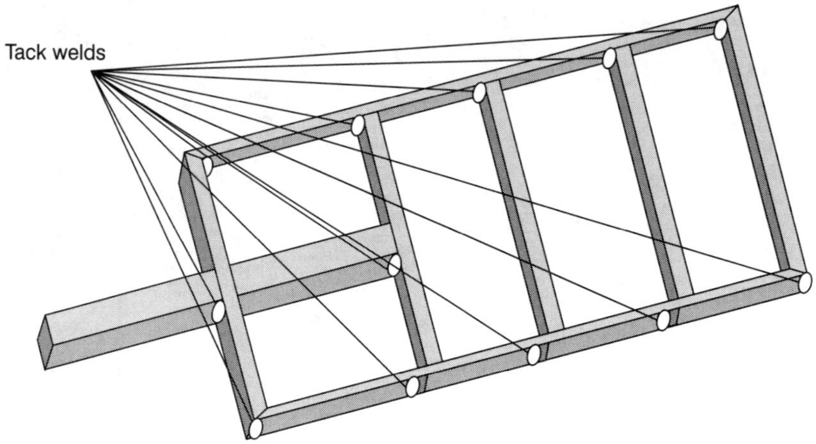

FIGURE 9-7 Trailer assembly tack welded together before being welded.

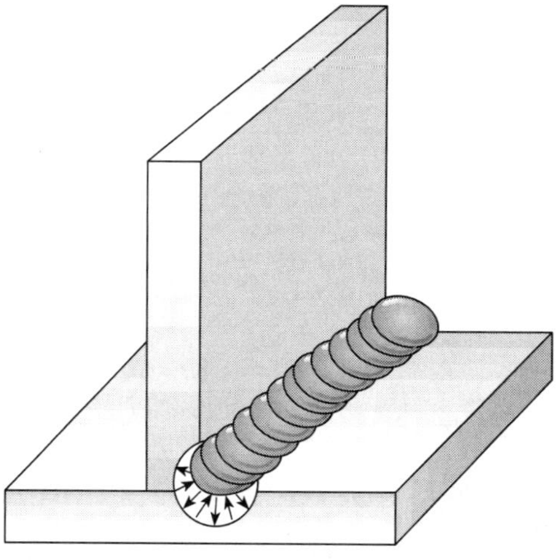

FIGURE 9-8 Welds shrink on cooling.

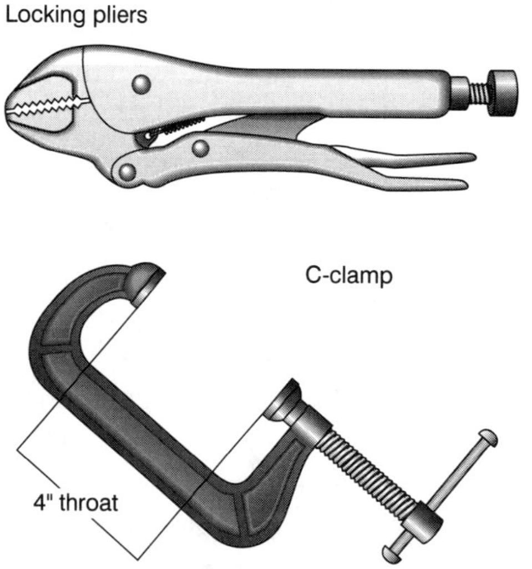

FIGURE 9-9 Handy tools for holding parts in position for welding.

Four 4-inch throated C-clamps and two locking pliers are the minimum amount of holding tools for any fabrication project (Figure 9-9). But if these tools are unavailable, don't be afraid to tack weld the project to a plate of steel. Anything that can be used should be used to hold the weldment in position to limit distortion. Remember distortion is a constant companion in welding that never goes away. Learn to limit the effects of distortion caused by welding, and even use distortion to your advantage.

FOR REVIEW

1. Why does distortion always occur as a result of welding?
2. What happens to metal as it is heated; as it cools?
3. What is the purpose of tack welds?

Lab Exercises

1. List the tools in your shop that can be used to control the distortion from welding.
2. Under supervision, use an oxyacetylene flame to straighten a crooked or bent steel rod.

REVIEW QUESTIONS FOR UNIT 9

Multiple Choice

Choose the *best* answer to complete the statement.

1. Low-alloy steels:
   a. Are the same as low-carbon steels.
   b. Are found in mining and agriculture.
   c. Require precautions when welding.
   d. Do not require preheating.
   e. Are answers a. and c. above.
   f. Are answers b. and d. above.
2. Martensite:
   a. Forms in HAZ.
   b. Forms when steel cools rapidly.
   c. Can lead to cracking in the joint.
   d. Is all of the above.
   e. Is a. and b. above.
3. Metal:
   a. Contracts when heated.
   b. Will return to original shape if restricted.
   c. Contracts on cooling.
   d. Is all of the above.
   e. Is b. and c. above.
4. Distortion:
   a. Results from uneven cooling.
   b. Always occurs in welding to some degree.
   c. Can be limited by tack welds.
   d. Is all of the above.
   e. Is b. and c. above.
5. Metal with the highest coefficient of expansion is:
   a. Copper.
   b. Steel.
   c. Aluminum.

d. Cast iron.
e. Stainless steel.

6. To limit stress:
a. Increase heat input.
b. Use unequal fillet welds in most applications.
c. Do no more welding than necessary.
d. Do all of the above.
e. Do none of the above.

7. Steel:
a. Expands twice the rate of aluminum.
b. Is less readily subject to distortion than aluminum.
c. Is lighter than aluminum.
d. Is a. and b. above.
e. Is none of the above.

8. Heat:
a. Contracts metal.
b. Has greater input with larger electrodes.
c. Expands metal.
d. Is b. and c. above.
e. Is none of the above.

9. Thickness of the base metal:
a. Has little to do with the amount of welding.
b. Does not affect heat input.
c. Usually determines size of the weld.
d. Is all of the above.
e. Is none of the above.

Short Answer

1. Tell which has a greater coefficient of linear expansion, steel or aluminum steel or stainless steel?
2. Describe the relationship between the size of the weld and distortion.
3. Define *stress.*
4. Explain how the size of an unequal leg fillet weld is determined.
5. Explain how the effective throat is measured.
6. Explain what happens to the weld pool as it cools.
7. Explain how to keep tack welds from pulling.
8. Explain the effects of using larger electrodes when welding.
9. Explain why a ½-inch fillet weld is preferred over a 1-inch fillet weld.
10. Describe the similarities between a C-clamp and locking pliers.

ADDITIONAL ACTIVITIES

Word Find

Directions: Circle the technical terms listed in the letter grid.

A	E	T	A	L	P	O	S	T	H	E	A	T	ALUMINUM
E	L	T	T	L	O	E	L	I	T	C	U	D	DISTORTION
T	D	U	A	N	N	O	I	N	A	A	E	C	STRESS
I	I	A	M	C	O	T	R	N	I	D	M	K	STEEL
S	U	R	D	I	S	T	O	R	T	I	O	N	TACK WELD
L	M	T	T	T	N	K	T	T	A	E	L	T	WELD ROOT
E	C	N	R	Z	S	U	A	S	C	E	E	N	PLATE
E	A	E	O	D	N	E	M	O	K	P	W	E	JOINT
T	S	C	P	G	H	E	R	E	W	D	P	M	WELDMENT
S	B	T	A	E	S	T	I	S	E	M	A	D	POSTHEAT
M	O	E	R	S	I	A	T	T	L	N	P	L	DUCTILE
N	O	I	S	N	A	P	X	E	D	N	G	E	EXPANSION
Y	J	O	I	N	T	O	O	R	D	L	E	W	

Writing Skills

1. In a paragraph, describe some properties in metal.
2. In a paragraph, define *martensite* and explain how to prevent its formation as a result of welding.
3. In a paragraph, explain how to prevent distortion.

Math

Decimals

0.1 = tenth
0.01 = hundredth
0.001 = thousandth
0.0001 = ten-thousandth
0.00001 = hundred-thousandth
0.000001 = millionth

Addition examples: $0.1 + 0.001 + 0.01 = 0.111$
$0.0001 + 0.00001 + 0.000001 = 0.000111$

1. $0.125 + 2 + 0.45 = ?$ **2.** $0.025 + 0.0044 + 0.347 + 0.0178 = ?$

3. $0.00051 + 0.0000088 + 0.000009 = ?$

4. $0.0000123 + 0.0000088 + 0.000009 = ?$

Multiplication examples: $0.000002 \times 5 = 0.000010$
$0.0000025 \times 82 = 0.000205$ $0.000009 \times 20 = 0.00018$

5. $0.0000063 \times 7 = ?$ **6.** $0.0000123 \times 22 = ?$

7. $0.000009 \times 15 = ?$

The coefficient of linear expansion is a number given to a material to measure its expansion per inch per degree of rise in temperature. For example, if, in the morning a length of aluminum is measured at 16 feet, what is its length at noon with a 53° rise in temperature? 16×12 (12 inches per foot) $\times 53 \times 0.0000123 = 0.1251648$ or $\frac{1}{8}$ inch. New length 16′ $\frac{1}{8}$″.

coefficient of linear expansion: steel = 0.0000063 stainless steel = 0.000009 aluminum = 0.0000123

8. The steel beam which measured a length of 99 feet in the morning did not fit into place that afternoon. What was its length if its temperature jumped 40°?
9. The stainless steel container which measured 2-foot square did not fit into place. Having been heated, its temperature jumped 290°. What was the amount of increase in size?
10. The 60-foot aluminum boat had a panel cut for placement in the hull that was 1/16 inch too large and did not fit. If the panel's temperature was controlled and the temperature of the hull jumped 15°, would that be enough so the panel fit? Show your work.

Puzzlers

1. A nut had seized on a ½-inch bolt. It was calculated the nut would have to be expanded three-thousandths of an inch to be removed. How hot would the nut have to be heated and what color would be visible when the nut reached the approximate temperature?
2. Does titanium have a greater coefficient of linear expansion than steel? Explain your answer.
3. An aluminum bar 12 feet long would have to experience a rise of 71° before its length would increase ⅛ inch. How long would a length of steel have to be to experience the same increase of length at 71° (rounded off to the nearest foot)?

UNIT 10

The Welding Project

REPAIR AND FABRICATION

Repair

The welding skills required to meet different job opportunities will depend on the amount of training a welder has received. Training can vary from several hours to several years. Job levels and the pay offered are determined by the skill of the welder. Regardless of the job and the qualification test, all welding will be either repair or fabrication.

So many things are discarded daily that could have been repaired by welding. Welding is one tool that can be used to recycle products that might otherwise be thrown out. Welding can contribute in a positive way from what has become a throwaway society. A welder can save energy, save resources, and when the employer is a large company, save money by doing repairs on things that wear out or break. A leak in a pipeline, the wear in a shovel bucket, or a break on a piece of earthmoving equipment can often be repaired (Figure 10-1). Much is gained by repairing things instead of throwing them away; in some cases there is no alternative.

Not everything repairable by welding can be fixed by the gas metal or the flux cored arc welding process. Sometimes things break under stress, wear down, or simply rust away. Because the failure of equipment and its parts is an ongoing problem, the repair of metal parts will always require the service of welders. Repairs, however, are not limited to industrial products—home lawn and garden tools can be repaired or redesigned by welding. The cross members of a snowmobile trailer subjected to the corrosion of road salt can be easily replaced by welding without changing the design of the trailer (Figure 10-2).

In the past, manufacturers constructed overly heavy equipment. Machinery was reinforced with lots of steel. Today's engineering is

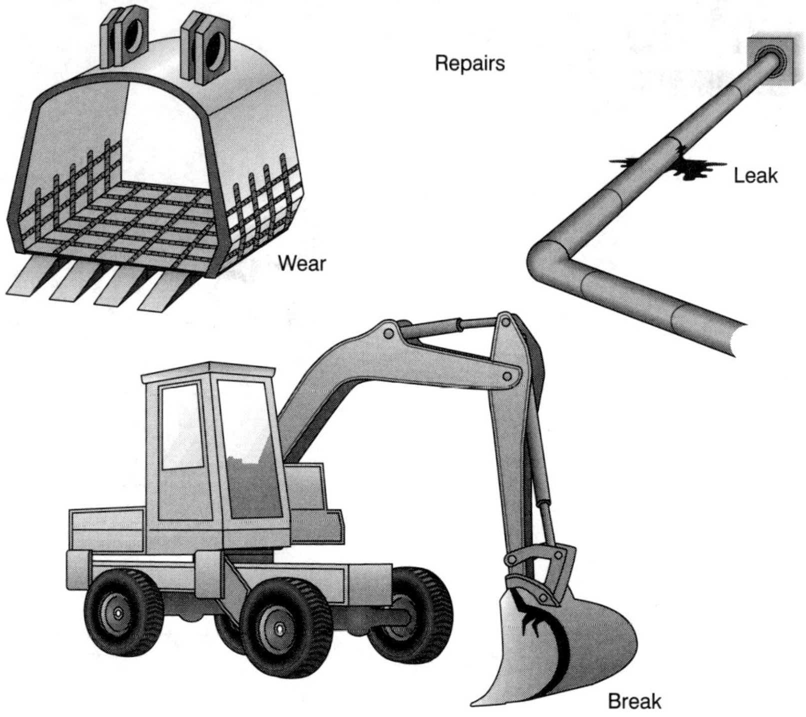

FIGURE 10-1 Repair by welding can take many forms.

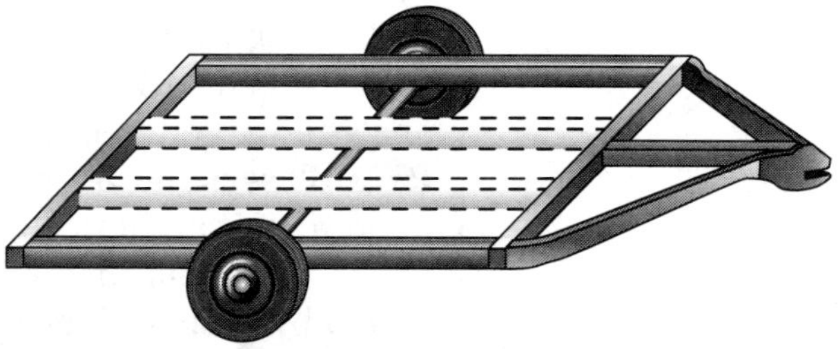

FIGURE 10-2 Trailer rebuilt by welding.

better at measuring the forces acting on metal. This knowledge has been useful in the manufacturing of products made of lighter weight design. For example, in the automobile industry lighter vehicles produce better gas mileage, but create problems that must be solved to prevent personal injuries of the passengers. Thus, new challenges are created by the use of lighter and more corrosion-resistant metals, with solutions still waiting to be found for the problems these new ideas create. For example, some lighter weight metals are more difficult to weld consistently in high volume manufacturing processes.

Welding has a major role in the repair of costly equipment by extending its serviceable life. There is no need to discard an otherwise functional tool or piece of machinery that can be repaired by welding.

Fabrication

Fabrication can involve the copying, modification, or creation of entirely new products. Taking an idea and turning it into a useful new product can require welding. Applying professional skills, the welder can transform a few shapes of cold, lifeless metal into

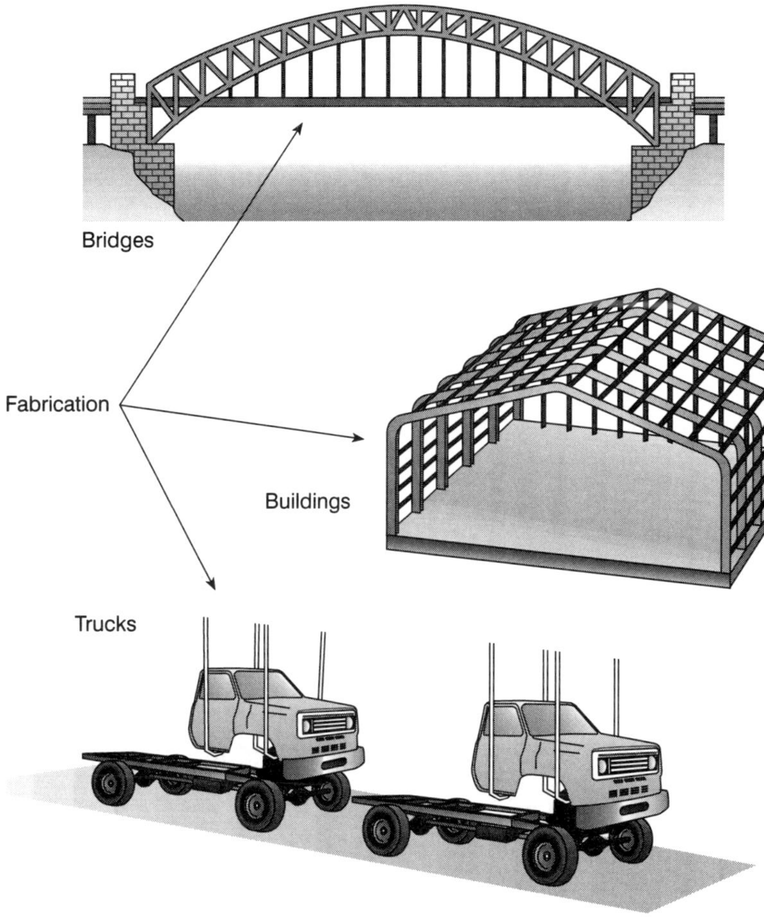

FIGURE 10-3 Commercial fabrication by welding provides many jobs.

something that is both strong and practical. Constructing bridges, trucks, and buildings requires the employment of many welders (Figure 10-3).

There are thousands of things around the home that can be made using welding equipment (Figure 10-4). A project may require new material, but junked metal from discarded products should not be overlooked. So often the rust is only on the surface and can be removed quite easily by grinding.

Refer to Unit 1 and the variety of structural shapes discussed there. A large metal supplier can provide a materials handbook for all the shapes they carry. Note how structural shapes are measured in Figures 10-5, 10-6, and 10-7. When ordering **pipe,** for example, state *the number of lengths by the inside diameter by the length.* Pipe up to and including 12 inches is measured by the inside diameter. Buy **hot-rolled steel** which is less expensive, not having undergone the extra rolling process that adds to its strength.

Tools

Some tools provide assistance to keep a project in position without moving. Some of the handy tools for holding a weldment in place for welding are the C-clamp, vice grips, and the vice. These tools also prevent the weldment from being pulled out of shape as

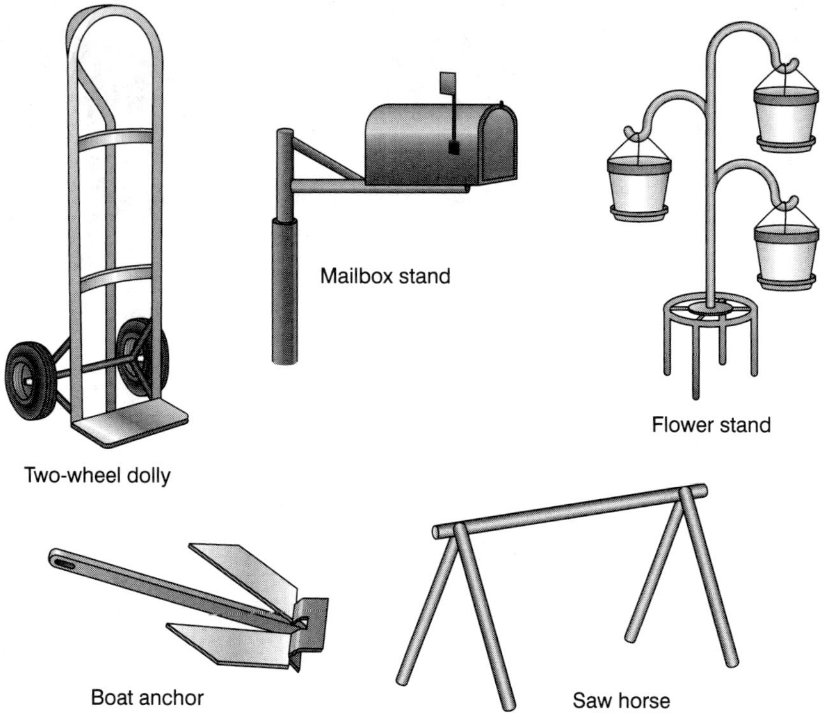

FIGURE 10-4 Practical fabrication projects for the home workshop.

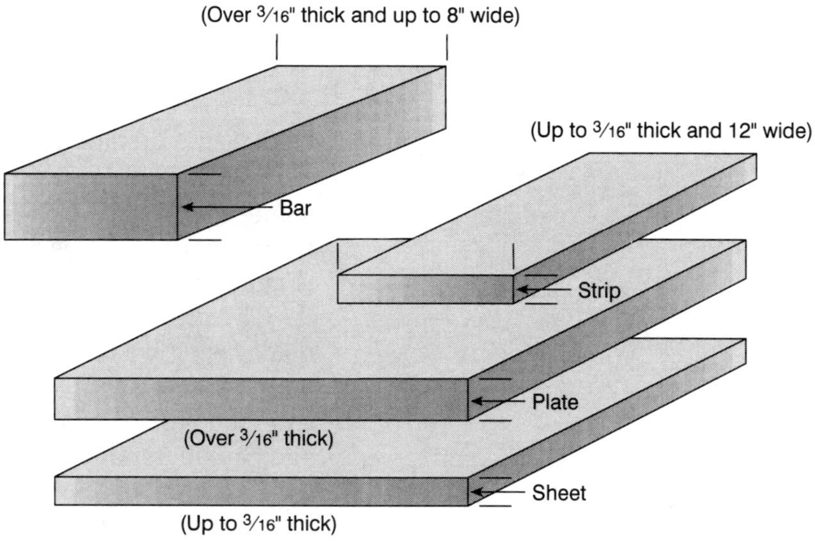

FIGURE 10-5 Structural shapes: bar, strip, plate, and sheet.

the welds cool and contract. Because welds rust quickly when exposed to moisture, a coat of primer designed for metal offers protection. If a second coat of paint is added over the primer, this will give the weldment a finished appearance.

An electric grinder is indispensable in any shop where welding is being done. Rusted metal or metal with its bluish mill scale covering can be cleaned by grinding or sandblasting before welding. Mill scale, rust, and old welds can be removed quickly by grinding. The grinder also has application for minor jobs that require cutting metal or making a bevel on the edge. The proper technique for us-

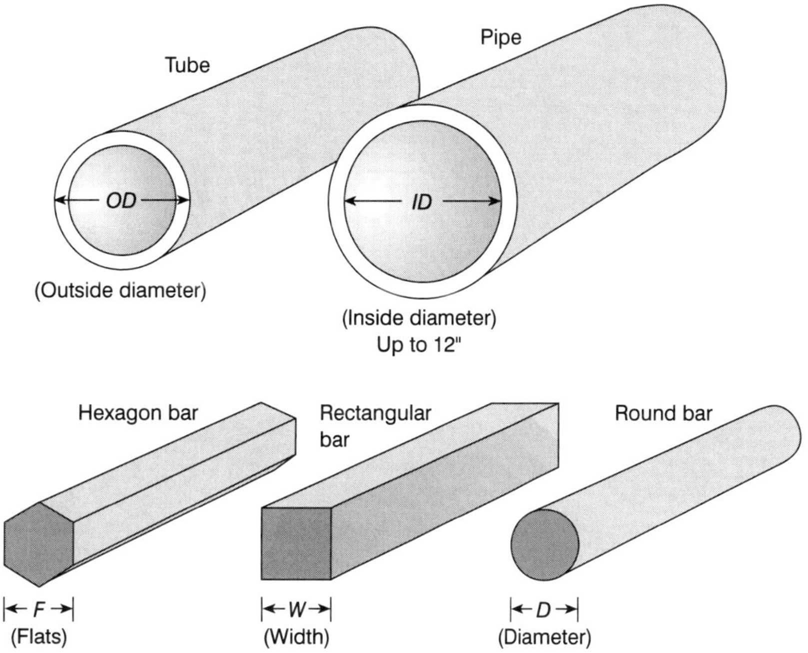

FIGURE 10-6 Structural shapes: tube, pipe, and bar.

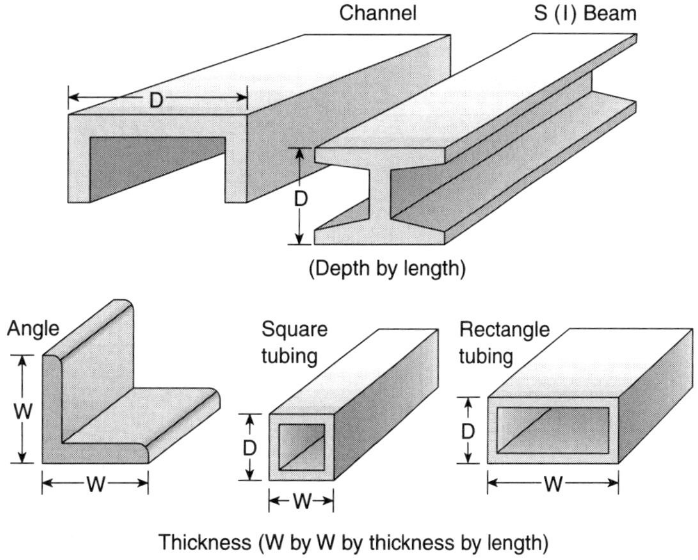

FIGURE 10-7 Structural shapes: angle, beam, channel, and tubing.

ing a grinder calls for tilting the grinder so that the bottom of the grinding disk makes contact with the weldment just below the edge (Figure 10-8). Do not use the grinder as you would use a polishing tool where the entire bottom surface of the disk makes contact.

When welding an item in which the alignment is critical, tack weld the broken part together and then remove necessary metal in the crack before welding (Figure 10-9). In this way, when the part is welded back together, it will be aligned. Be careful if making a bevel on a thick part that was previously broken not to bevel the entire edge, otherwise the part may not fit back together as originally manufactured.

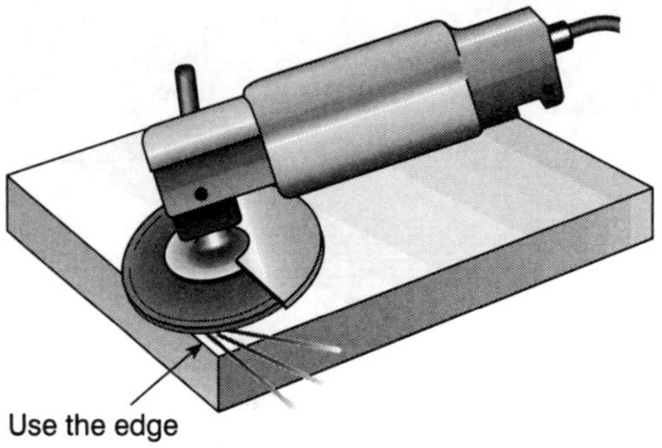

FIGURE 10-8 Proper use of grinder requires contact of the edge on the grinding disk, not the entire bottom surface.

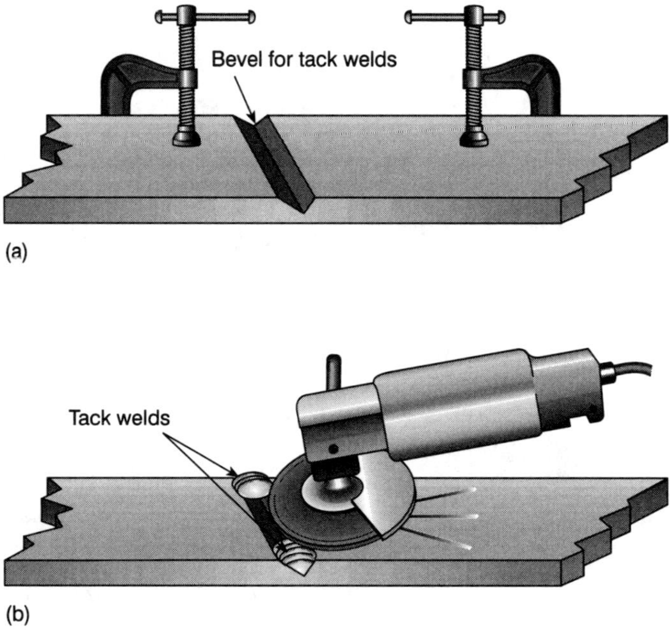

FIGURE 10-9 (a) Position part, clamp, and bevel. (b) Tack weld and bevel crack before completing the weld.

Other tools to be considered for any shop are a jack, cleats, and wedges (Figure 10-10). A hydraulic jack may be necessary in repair for bringing parts together. Cleats and wedges can be fabricated for use in realigning parts. Fittings can be fabricated so a jack will work to apply a controlled force (Figure 10-11). Other useful tools include a steel square, a soapstone, a combination square, and a punch (Figure 10-12). Helpful electric tools include the drill, metal cutting saw, and shears (Figure 10-13).

Layout

When laying out patterns for cutting on sheet or plate, try to work off a straight, mill-cut edge (Figure 10-14). Check the edge with a framing square or combination square to keep measurements accurate. Then after measuring and marking with soapstone, make your cut on the line or outside the line. Cutting outside the line will be more accurate, because grinding the edge will remove any rough spots, bringing the cut back to the measurement as marked.

ARC FLASH

Make a freehand sketch of any fabrication project to help visualize the appearance of the finished product.

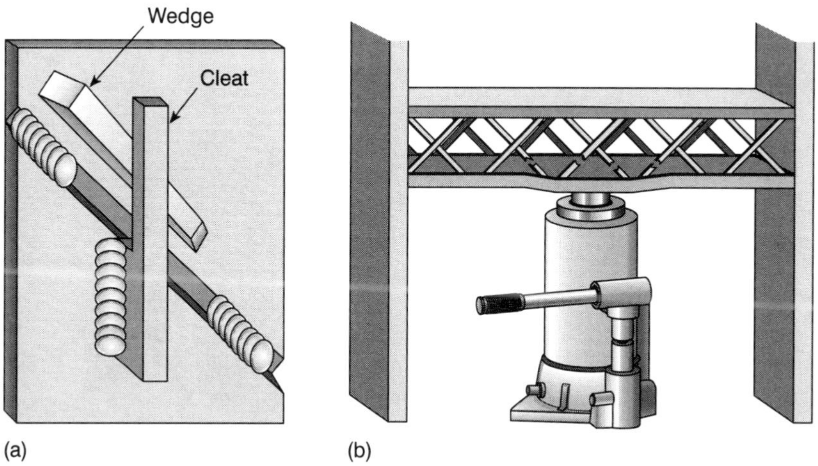

FIGURE 10-10 (a) Use of wedge and cleat for realignment. (b) Use of jack for realignment.

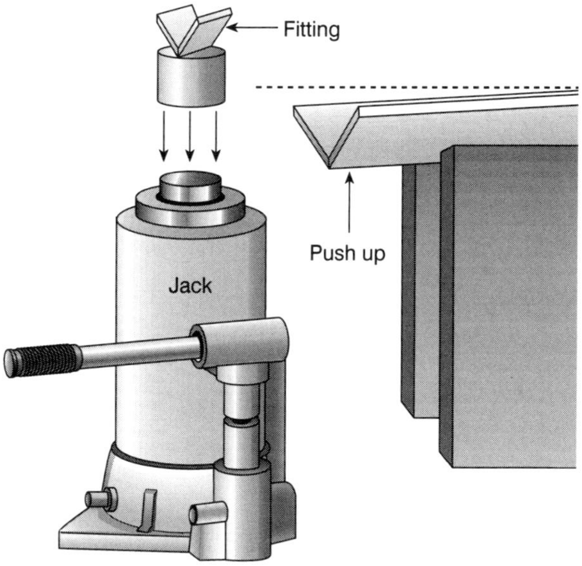

FIGURE 10-11 Use of fabricated fitting for assistance.

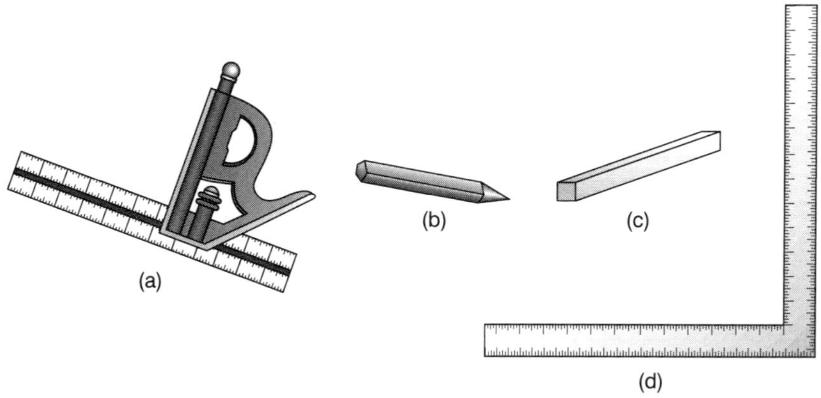

FIGURE 10-12 Useful tools: (a) combination square; (b) punch; (c) soapstone; and (d) steel square.

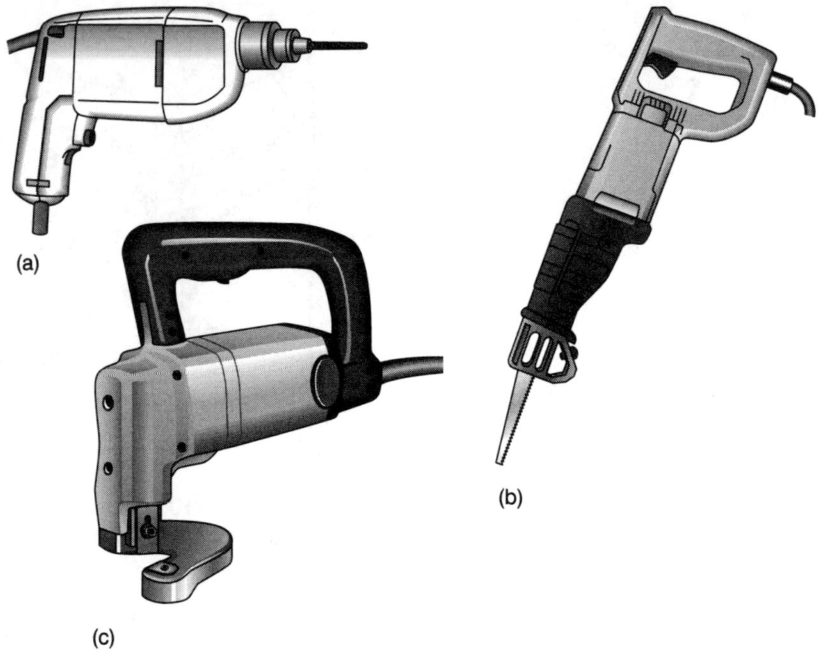

FIGURE 10-13 Useful power tools: (a) electric drill; (b) electric cutting saw; (c) sheet metal shear.

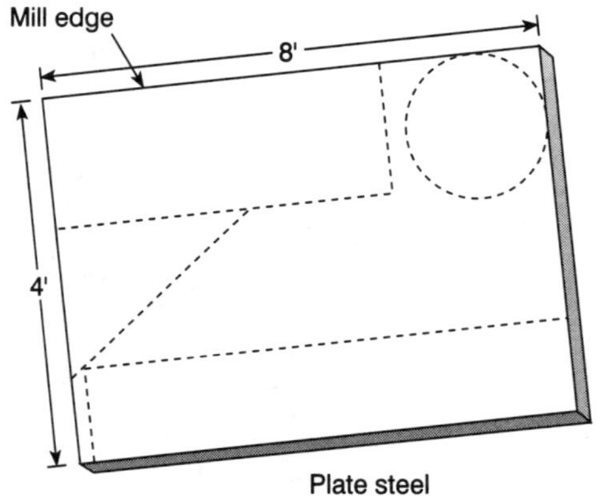

Figure 10-14 Organize shapes when cutting to conserve metal.

After completing the welding, set the weldment aside. Come back later with a fresh set of eyes. Are all the required welds in place? If anything is wrong or out of place, this second look may catch the mistake.

SAFETY REMINDER

Be sure paints, thinners, and any other volatile or flammable materials are kept out of the shop area.

Lab Exercises

1. Repair by welding a project.
2. Sketch out (computer design) a project you would like to construct by use of welding.
3. Fabricate a project by welding.

Appendix

GUIDED BEND TEST

Introduction

The **guided bend test** is a method for examining a welder's ability to follow a given procedure. This is a **destructive test** in which **coupons** are bent, stretching to examine for flaws in the welding. The guided bend test is used for making **face bends, root bends,** or **side bends.** The face bend stretches the weld face. The root bend stretches the root. The side bend stretches the side (Figure A-1).

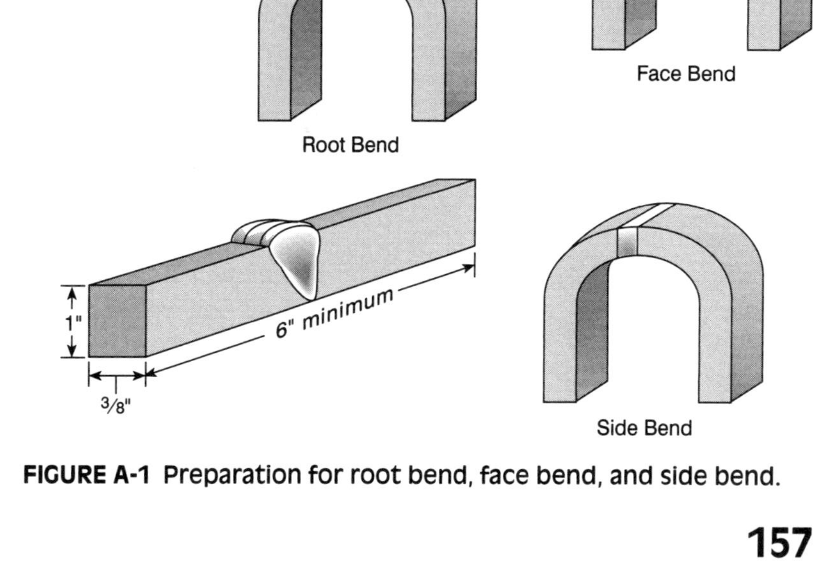

FIGURE A-1 Preparation for root bend, face bend, and side bend.

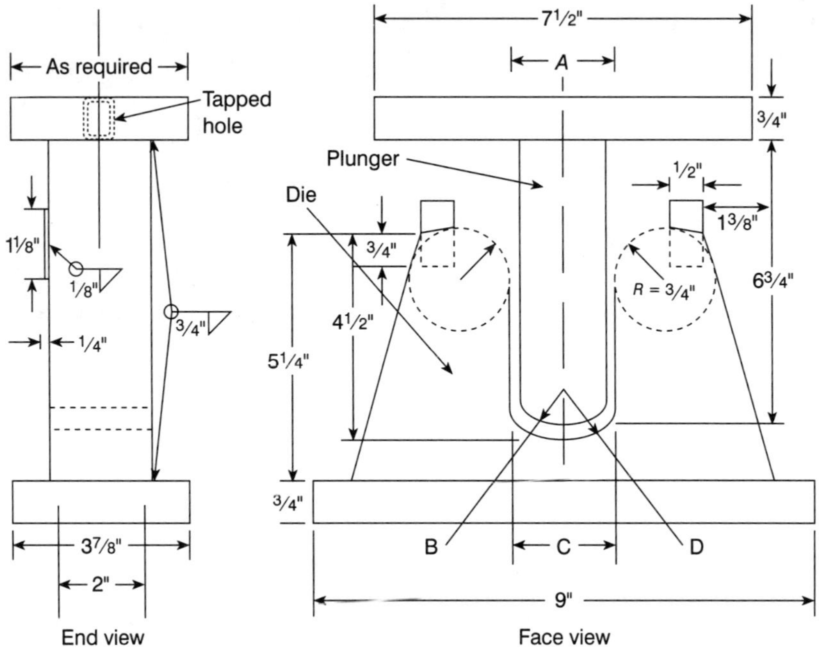

FIGURE A-2 Jig design.

TABLE A-1 Dimensions for three jigs

Yield Strength (psi)	*A*	*B*	*C*	*D*
50,000 and under	1½ inch	¾ inch	2⅜ inch	1³⁄₁₆ inch
50,000 to 90,000	2 inch	1 inch	2⅞ inch	1⁷⁄₁₇ inch
Greater than 90,000	2½ inch	1¼ inch	3⅜ inch	1¹¹⁄₁₆ inch

The guided bend test requires a jig for bending the coupons. The design for the jig illustrated in Figure A-2 is specified by the American Welding Society's *D1.1, Structural Welding Code-Steel* (Table A-1). A hydraulic jack is used to push either the shoulder or the plunger onto the coupon, bending and stretching the material. A36 steel has a yield strength of 36,000 psi minimum.

Qualifications of Testing

The specifications are given for testing groove welds of limited thickness (Figure A-3) and unlimited thickness (Figure A-4). The groove weld test of the butt joint, according to the *D1.1 Structural Welding Code-Steel*, can be given in any of the four positions for welding, with the limitations dependent on the thickness of the base metal (Table A-2). Successful completion of tests in the vertical and overhead positions qualifies a welder in all four positions.

A36 steel

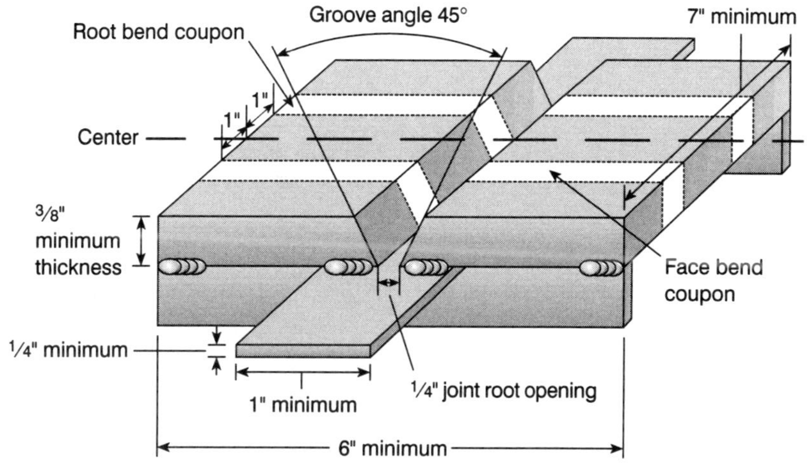

FIGURE A-3 Specifications for groove welds of limited thickness.

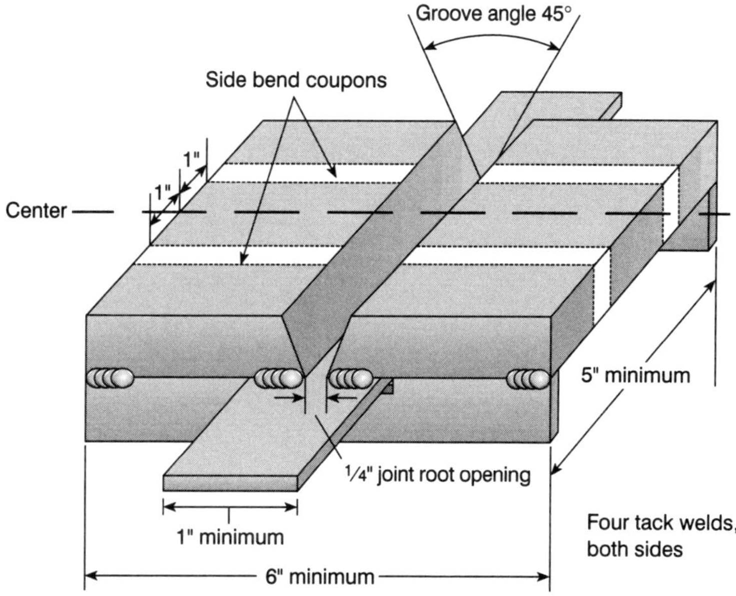

FIGURE A-4 Specifications for groove welds of unlimited thickness.

Preparing Coupons for Testing

Coupons must be taken from designated areas in the weldment as noted in the specifications given in Figures A-3 and A-4. A jig designed for 50,000 psi or less meets the requirements of A36 steel with a yield strength of 36,000 psi minimum.

In preparing coupons for testing, they can be cut to size with a band saw. Reinforcement on the weld face, root face, backing strip, and strong backs must be removed; this can be done with a torch. Grind the weld flush using a hand grinder or a pedestal grinder, leaving the grinding marks parallel to the length of the coupon. File the edges of the coupon to a ⅛-inch maximum radius (Figure A-5).

TABLE A-2 Welding limits for successful testing ⅜-inch groove welds.

Position Tested	*Type of Weld*	*Position Qualified*	*Thickness Qualified*
Flat	Groove Weld	flat	Up to ¾"
Horizontal	Groove Weld	flat and horizontal	Up to ¾"
Vertical	Groove Weld	flat, horizontal, and vertical	Up to ¾"
Overhead	Groove Weld	flat and overhead	Up to ¾"
Welding Limits for Successful Testing 1-inch Groove Welds			
Flat	Groove Weld	flat	Unlimited
Horizontal	Groove Weld	flat and horizontal	Unlimited
Vertical	Groove Weld	flat, horizontal, and vertical	Unlimited
Overhead	Groove Weld	flat and overhead	Unlimited

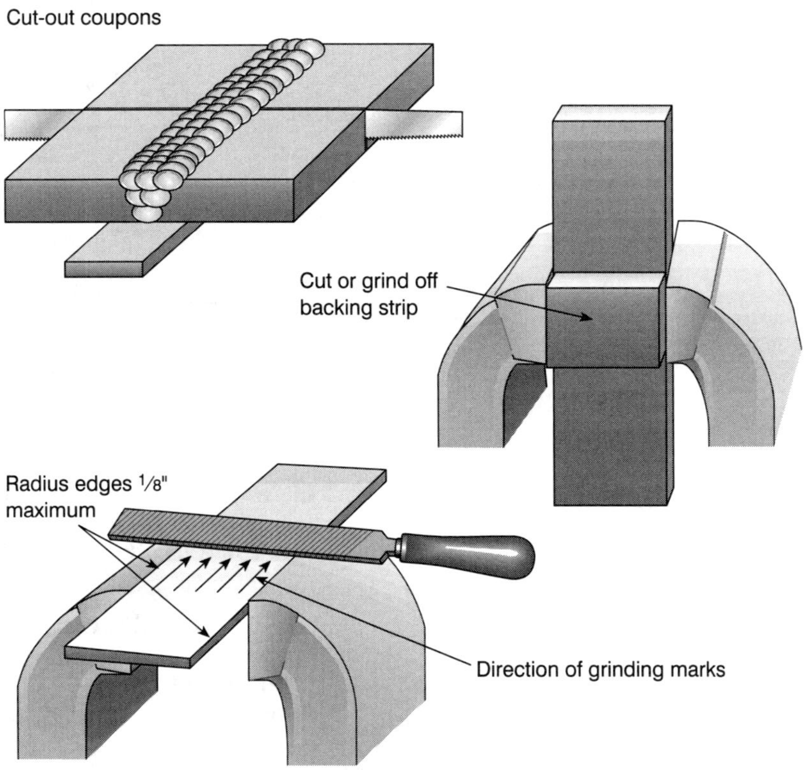

FIGURE A-5 Preparation of coupon for testing.

Testing the Coupons

The welder shall have passed the *D1.1 AWS Structural Welding Code-Steel* qualification for guided bend test if the coupon meets the following criteria:

1. No **discontinuities** of ⅛ inch or more in length, measured in any direction.
2. Sum of all discontinuities shall not equal or exceed ⅜ inch.
3. A crack along the edges of the coupon that contain no visible slag shall measure no more than ¼ inch.
4. A crack along the edges of the coupon that were filed and that contain visible slag shall measure no more than ⅛ inch.
5. A coupon with corner cracks greater than ¼ inch shall be replaced with another test coupon from the original weldment.

Passing a guided bend test supervised by a licensed testing faculty will be required, and will then provide the welder proof of the accomplishment. Many companies require that their welders pass both written and destructive tests before hiring.

LAB EXERCISE ONE

Short Circuiting Transfer (GMAW-S): Tee Joint in Vertical Position

Fillet Weld

Position the gun correctly, using the proper work angle and travel angle. Be sure to move in a steady motion, pausing from side to side so that the weld bead is equally deposited on both pieces of the joint. Make several passes, alternating from each side of the joint (Figure A-6).

Equipment and Material

Safety glasses
Helmet
Earplugs
Grinder
Pieces of ¼- to ⅜-inch steel, 3″ × 8″ (2)
Welding gloves
Protective clothing
Sidecutter
Spool wire 0.035 (ER70S-3)
Shielding gas: carbon dioxide or 75% argon to 25% carbon dioxide
C-clamp fixture

Welding Parameters

Amperes: 140 to 250
Voltage: 19 to 24
Electrode extension: ¼ to ⅜ inch

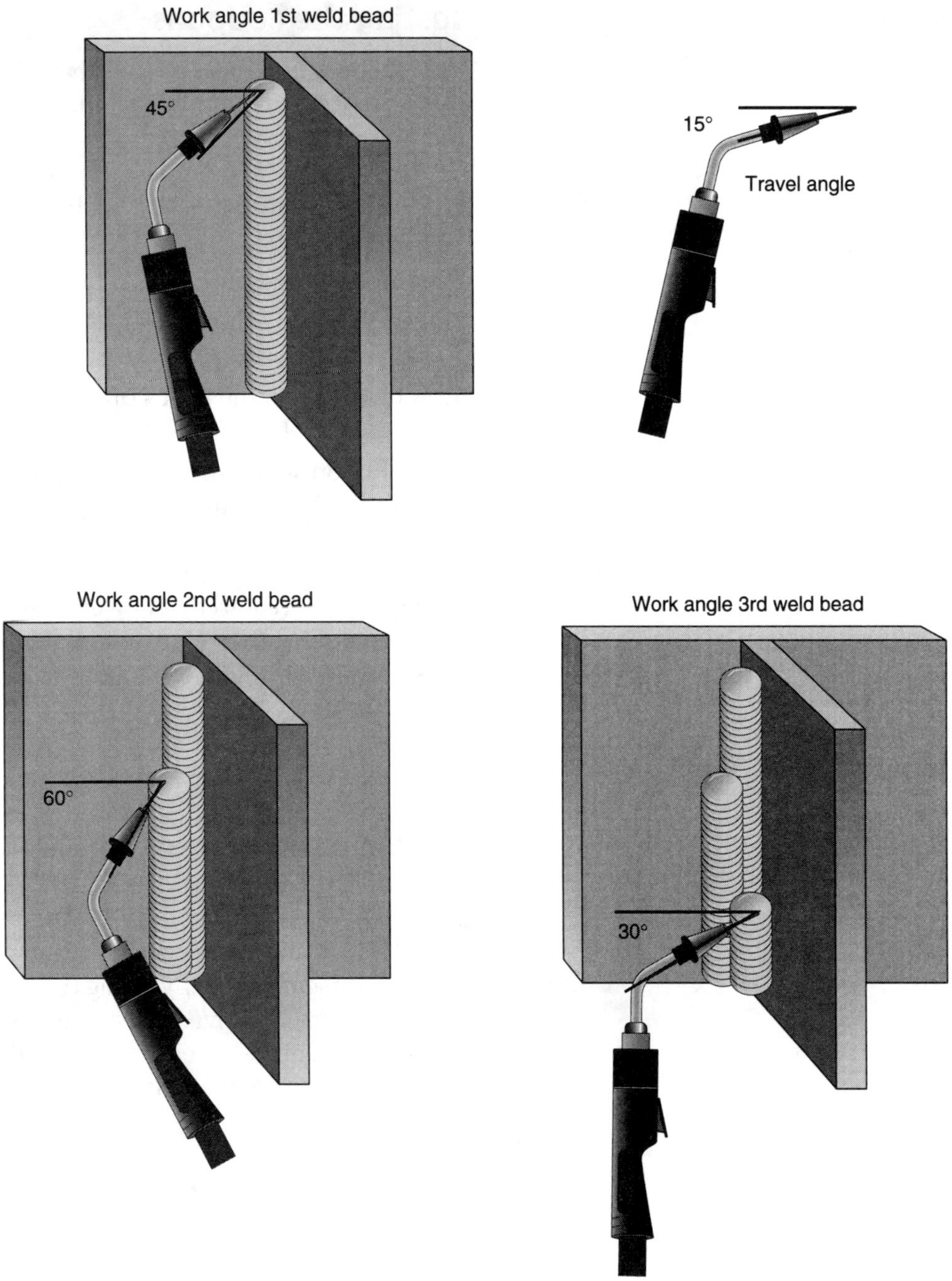

FIGURE A-6 Cool the joint when necessary.

Directions

1. Clean the surfaces of the base metal with the grinder.
2. Make sure the power is off before plugging in the power source. Switch on power.
3. Adjust the gas flow from 15 to 25 cfh.
4. Set the voltage and set the wire speed at 50%, adjust by running test runs. Observe voltmeter and ammeter during welding to adjust the welding parameters.
5. Tack weld the pieces of the joint together. Position the joint in the vertical position, using a C-clamp fixture, etc.

6. Begin at the bottom to make the first pass, then alternate to the other side.
7. Make a visual inspection. Look for smooth even weld, distributed equally on both sides of the joint with no porosity.
8. Check with the instructor for evaluation, if necessary, before proceeding to the next exercise.

LAB EXERCISE TWO

Spray-Arc Transfer (GMAW-SP): Butt Joint Meets AWS Structural Welding Code

V-Groove Weld with Backing in Vertical Position

Exercise Two is designed for gas shielding (can be completed using the flux cored arc welding process). Evaluate with a destructive test. Note that a backing strip is used in this exercise. Fit-up is important, so make sure the dimensions are accurate. Go over the tack welds when completing the first pass (Figure A-7).

Equipment and Material

Safety glasses
Helmet
Earplugs
Grinder
Pieces of ⅜-inch A36 steel, 3″ × 7″ (2)
Welding gloves
Protective clothing
Sidecutter
2″ × 10″ (backing)
Strong backs (2)
Spool wire 0.045 (ER70S-3)
Shielding gas: 98% argon to 2% oxygen
Cutting machine, 22.5° bevel

Welding Parameters

Amperes: 180 to 225
Voltage: 23 to 26
Electrode extension: ⅜ to ¾ inch

Directions

1. Clean the surfaces of the base metal with the grinder.
2. Make sure the power is off before plugging in the power source. Switch on power.
3. Adjust the gas flow from 40 to 50 cfh.
4. Set the voltage and set the wire speed at 50%, adjust by running test runs. Observe voltmeter and ammeter during welding to adjust the welding parameters.
5. Tack weld the two pieces to the backing plate. Set a ¼-inch joint root opening. Tack weld the strong backs in four places.

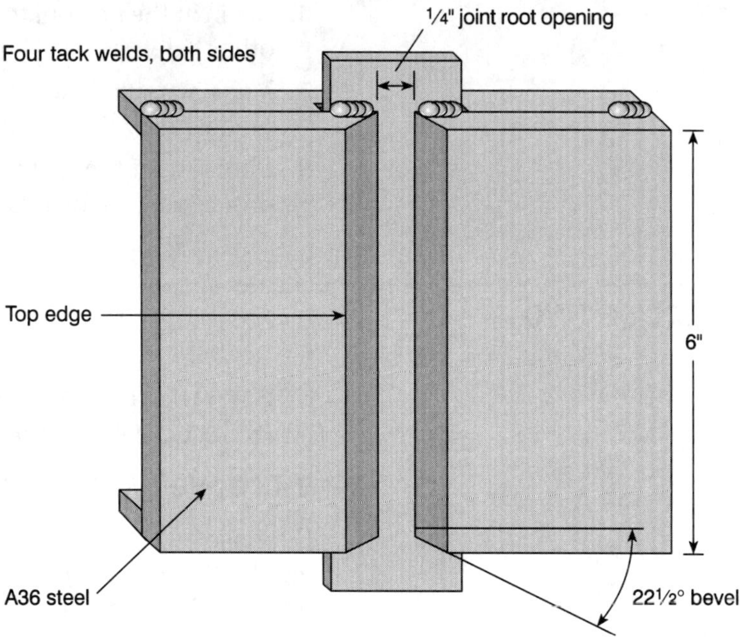

FIGURE A-7 Adjust the travel speed if necessary.

6. Make a sideward motion, making sure there is complete fusion on the joint root pass.
7. Make a visual inspection. Look for complete fill within the joint.
8. Check with the instructor for evaluation, if necessary, before proceeding to the next exercise.
9. Pass the guided bend test before proceeding.

LAB EXERCISE THREE

Short Circuiting Transfer (GMAW-S): Tee Joint in Overhead Position

Fillet Weld

Once again the position of the gun is important; use the proper work angle and travel angle. Be sure to move in a steady motion, pausing from side to side so that the weld bead is equally deposited on both pieces of the joint. Watch so there is no undercut. Make several passes, alternating from each side of the joint (Figure A-8).

Equipment and Material

Safety glasses	Grinder
Helmet	Pieces of ¼- to ⅜-inch steel, 3″ × 8″ (2)
Earplugs	

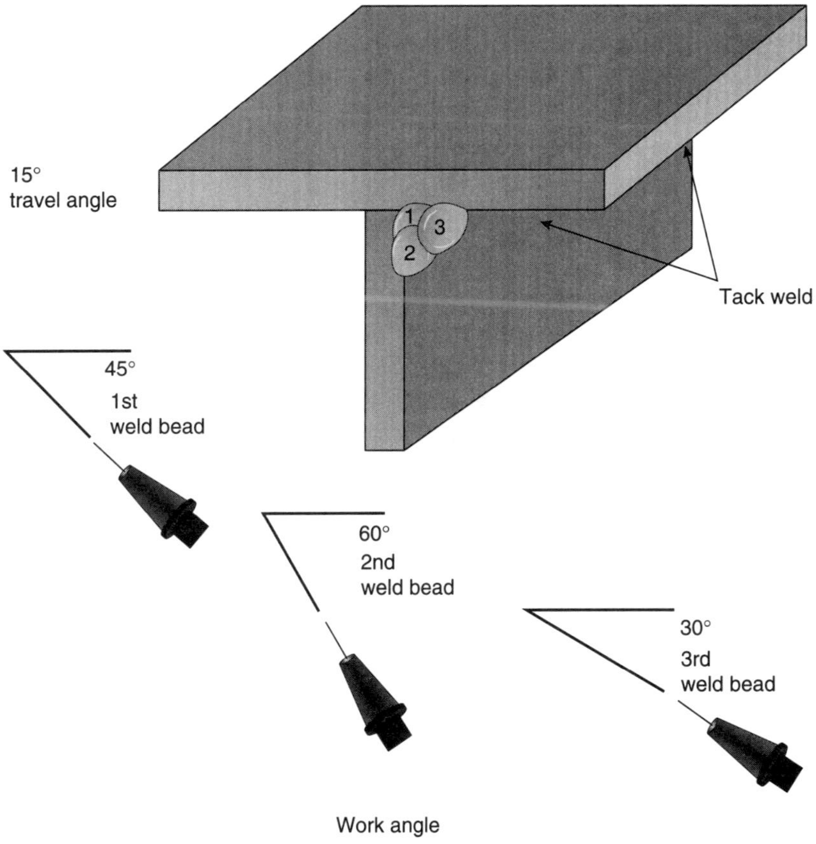

FIGURE A-8 Find comfortable position for welding.

Welding gloves

Protective clothing

Sidecutter

Spool wire 0.035 (ER70S-3)

Shielding gas: carbon dioxide or 75% argon to 25% carbon dioxide

C-clamp fixture

Welding Parameters

Amperes: 140 to 250

Voltage: 19 to 24

Electrode extension: ¼ to ⅜ inch

Directions

1. Clean the surfaces of the base metal with the grinder.
2. Make sure the power is off before plugging in the power source. Switch on power.
3. Adjust the gas flow from 15 to 25 cfh.
4. Set the voltage and set the wire speed at 50%, adjust by running test runs. Observe voltmeter and ammeter during welding to adjust the welding parameters.
5. Tack weld the pieces of the joint together. Position the joint in the overhead position using a C-clamp fixture, etc.

6. Begin at the bottom to make the first pass, then alternate to the other side.
7. Make a visual inspection. Look for smooth even weld, distributed equally on both sides of the joint with no porosity.
8. Check with the instructor for evaluation, if necessary, before proceeding to the next exercise.

Glossary

AC—See *alternating current.*

active gas—A gas such as carbon dioxide that forms compounds readily.

alloy steel— A steel in which one of the following amounts is exceeded: 1.65% manganese, 0.60% silicon, or 0.60% copper; or, a steel to which a definite minimum quality of aluminum, boron, chromium, or any other alloying element is added to achieve a desired alloying effect.

alternating current (AC)—The back-and-forth motion of electric current, changing directions 120 times (60 cycles) per second.

aluminum—A popular nonferrous metal that has one-third the weight of steel.

American Iron and Steel Institute—An organization that classifies types of steel.

American Society for Testing and Materials (ASTM)—A specification-forming body that helps to establish quality through standardized testing.

American Welding Society (AWS)—An organization that plays a major role in setting standards used throughout the welding industry.

amperage—(1) A measurement in the rate of flow; (2) wire speed setting on a power source.

arc—See *welding arc.*

arc blow—An erratic arc that refuses to go where the welder desires and results in sputter and spatter.

arc flash—A painful, but usually temporary, eye condition caused by the light of the welding arc on unprotected eyes.

arc length—The distance of the electrode from the weld pool.

arc strike—The beginning of the arc for welding. An arc strike can lower the weld quality if made on the base metal outside of the welding area.

argon—An inert shielding gas used in some arc welding processes.

automatic—A term used to describe welding independently with only occasional adjustment of controls.

AWS—See *American Welding Society.*

backhand—A welding technique in which the gun nozzle is focused on the weld pool opposite the direction of travel.

backing—A material commonly used with groove welds on the backside of the joint.

base metal—The material being welded.

bevel—An edge preparation of the base metal for welding in which any angle less than 90° is formed.

bevel-groove weld—The use of the butt joint with edge preparation given to one of two pieces forming the joint.

birdnesting—The result when wire becomes tangled in the drive rolls.

boron—A nonmetallic element used as an alloy to increase hardness.

brass—An alloy consisting of copper and zinc.

brazing—A group of joining processes in which the filler metal is heated to liquidus above 840°F but below the melting temperature of the base metal, and capillary action takes place.

buildup—A filler metal deposited so as to extend above the surface of the base metal to rebuild worn parts.

butt joint—A joint formed with pieces aligned on the same plane.

carbide precipitation—A chemical reaction of carbon with chromium that reduces the corrosion resistance of the stainless steel.

carbon—A nonmetallic element added to steel and cast iron that affects hardness and, consequently, strength.

carbon dioxide—An active shielding gas used in some welding processes.

carbon steel—A general term covering a large range of steels. See *low-*, *medium-*, *high-*, and *very high-carbon steel.*

cast iron—An alloy of iron that contains carbon (up to 4.5%).

chromium—An element added to steel to increase hardness at high temperatures and corrosion resistance. Chromium is the principal alloying ingredient of stainless steel.

coefficient of linear expansion—A number given to a material as a measurement of its expansion per inch per degree of rise in temperature.

complete joint penetration—Penetration through the entire thickness of the joint is 100%.

compression strength—The resistance to being crushed.

concave—Shape or contour of a weld bead with an inward or depressed surface.

conductor tube—Transmits current from the copper cable.

constant current—The electric output by which the welder can, to a limited degree, control voltage by raising or lowering the electrode.

constant potential—The electric output in which voltage remains constant to maintain a given arc length even though the distance of the gun to the base metal changes.

contact tube—The point of electric contact that energizes the wire electrode.

convex—Shape or contour of a weld bead with a raised, rounded surface.

copper—A metallic element that is readily soldered, brazed, or welded.

corner joint—A joint formed with pieces aligned in different planes, involving the edge of at least one piece.

corrosion resistance—The ability to withstand the stress of chemical reactions without breaking down.

coupon—A dimensioned specimen from a weld or material for a destructive test to determine quality.

crack(ing)—A fracture of the joint that is the result of welding.

crater—An undesirable depression left when a weld bead is stopped or completed but not filled in.

crater crack—A crack originating in a crater.

current—The flow or movement of electricity.

DC—See *direct current.*

DCEN—See *direct current electrode negative.*

DCEP—See *direct current electrode positive.*

defect—A flaw in a weld as a result of welding that can affect weld quality.

deoxidizer—An element added to molten metal for removing oxygen.

destructive test(ing)—A test of weld quality in which the joint is destroyed.

direct current (DC)—Electric current that flows in one direction.

direct current electrode negative (DCEN)—Welding with the electrode attached to the negative terminal and the workpiece connection attached to the positive terminal on the power source.

direct current electrode positive (DCEP)—Welding with the electrode attached to the positive terminal and the workpiece connection attached to the negative terminal on the power source.

discontinuity—A flaw that is not a defect unless it fails to meet the condition established by the standards or specifications.

distortion—A usually uncontrolled change in the shape of metal, resulting from the heat of welding.

downhill—Vertical position, downward welding with gravity.

ductility—The resistance to failure after stretching, twisting, or bending.

duty cycle—The recommended percentage of time that a given power source should be under the load of welding.

earplugs—Provide protection from damaging high frequency.

edge joint—A joint formed when the edges of two or more pieces are brought together parallel to one another.

effective throat—The shortest distance from the weld root to the weld face at the point of a flush contour.

electrode—A consumable or nonconsumable material through which electricity flows to create an arc.

electrode extension—The distance between the end of the contact tube and the tip of the unmelted wire.

element—Any one of more than 100 substances made up of one kind of atom.

face bend—Stretching the weld face of a coupon to test the weld.

fatigue strength—The resistance to changing forces or loads.

FCAW—See *flux cored arc welding.*

feeder system—A part of the welding process that contains the drive rolls and controls for semiautomatic welding.

ferrite—The formation of iron in the grain structure of steel and cast iron.

ferrous—Containing iron.

filler metal—Metal added to a joint.

fillet weld—A weld that joins the surfaces of two pieces, usually at right angles to each other.

flammable—Easily ignited.

flat bar—Metal more than $\frac{3}{16}$ inch thick and usually not exceeding 6 inches in width.

flat position—A position in which the plane of welding is from 0° to 15°.

flowmeter—A gauge that measures gas volume by its flow.

flush contour—A weld face that is a straight line from toe to toe.

flux—A material that when transformed to a gaseous shield, protects the weld pool and stabilizes the arc, making welding easier.

Flux helps filler metal to flow and removes impurities that can affect weld quality.

flux cored arc welding (FCAW)—A semiautomatic arc welding process in which a tubular wire, filled with flux, is fed continuously into the weld pool.

forehand—A welding technique in which the gun nozzle is focused on the weld pool in the direction of travel.

frequency—The number of electrical oscillations (cycles) per second.

fusion—The melting and mixing together of the pieces that form the base metal with or without filler metal.

gas diffuser—Disburses the shielding gas through the nozzle.

gas metal arc welding (GMAW)—A semiautomatic arc welding process in which wire is fed continuously into the weld pool.

globular transfer—A gas metal arc welding process in which a drop greater than the diameter of the wire is produced.

GMAW—See *gas metal arc welding.*

grain structure—Configuration of the crystal composition within the metal.

groove angle—The measurement in degrees for the bevel or vee made on the edge or edges of the metal in preparation for welding.

groove face—The surface where the pieces of the joint meet.

groove weld—Any one of eight joint designs for the butt joint.

guided bend test—A test in which coupons are bent, stretching to examine for flaws in the welding.

gun—A device that transfers electric output to the filler wire coming out of its nozzle; contains an on/off switch.

hardness—The resistance to penetration or denting.

heat-affected zone (HAZ)—The area along the weld that is affected by the heat of welding.

helium—An inert shielding gas used in some arc welding processes.

high-alloy steel—A steel in which chromium, manganese, or nickel content equals 12% or better.

high-carbon steel—A steel that contains 0.45% to 0.65% carbon.

horizontal position—A position in which the plane of welding is from 15° to 100°.

hot-rolled steel—A steel that has been deformed into shape at temperatures that allow course grains to recrystallize.

impact strength—The resistance to sudden force without fracture.

incomplete fusion—Lake of fusion between the weld bead and the base metal.

inductance—A fine adjustment on some power sources that can reduce spatter and, more important, regulate weld pool time by increasing or decreasing arc time.

inert gas—A term used to describe noble gases, such as argon and helium, that neither react with oxygen nor form any other compounds readily.

infrared—Invisible light absorbed by the body as heat, which can damage the eyes.

input current—High-voltage, low-amperage electrical power that flows into a power source.

insulator—A device that does not conduct electricity and keeps the electrode holder from becoming part of the arc and short-circuiting against the base metal.

intermittent welding—Welding in which the welds are spaced and not continuous over the entire joint.

interpass—Time between the laying of the next weld bead.

inverter—A lightweight power source that takes advantage of computer technology. Some are versatile enough to be connected to most electric input systems.

iron ore—The raw material used in the manufacture of cast irons and steels.

joint—The junction of two or more pieces joined by welding, brazing, or soldering.

joint root—The closest part of the pieces forming a joint.

joint root opening—A separation at the joint root, usually for greater penetration.

lap(ping)—To lap over. The degree one weld bead extends over another weld bead.

lap joint—A joint formed with two pieces extending over each other in parallel planes.

layer—Lapping weld beads on a metal surface.

lead—Cable extending from the power source to the workpiece connection or the electrode holder.

liner—A device that carries the wire from the drive rolls to the contact tip.

longitudinal cracking—A flaw that occurs in the weld face or the base metal parallel to the toes of the weld.

low-alloy steel—A steel in which chromium, manganese, or nickel content is less than 12%.

low-carbon steel—A steel that contains 0.1% (0.001) to 0.3% (0.003) carbon.

manganese—A metallic element added to steel to increase its hardness.

manual—A term used to describe welding, brazing, soldering, or cutting performed using handheld equipment.

martensite—A hard, pinlike grain structure that forms when steel cools too rapidly for the pearlite grain structure to form.

medium-carbon steel—A steel that contains 0.3% (0.003) to 0.45% (0.0045) carbon.

melt-thru—Uncontrolled overheating that results in an unexpected hole.

mild steel—See *low-carbon steel.*

mill scale—Oxides (blue-black in color) on the surface of steel that were formed during manufacturing and can affect weld quality.

molybdenum—A metallic element added to steel to increase its hardness and tensile strength.

motor generator—A power source that uses direct current as output for welding.

nickel—An alloying element that increases the hardness of steel.

nitrogen—An active gas that comprises 78% of the air, and is damaging to the weld.

nondestructive testing —A test of weld quality in which the joint is not destroyed.

nonferrous—Containing no iron.

nozzle—A device that fits on the end of the gun to focus gas flow.

output—Low-voltage, high-amperage electric power that flows into the welding arc from a power source.

overhead position—A position in which the plane of welding is from 0° to 80°.

overlap—Flow of weld beyond the weld toe or weld root over unmelted base metal.

oxyacetylene—A welding, cutting, or brazing process that requires oxygen and acetylene.

oxygen—An element used under pressure in combination with shielding gas(es); oxygen comprises up 21% of the air, and when not controlled, can lower weld quality.

parameters—Important welding variables.

pearlite—The formation of ferrite and cementite (hard iron carbide) in the grain structure of steel and cast iron.

pipe—A structural, steam- and fluid-carrying medium measured from inside diameter up to and including 12-inch diameter. Pipes over 12 inches are measured by outside diameter.

plate—Metal that is more than 3/16 inch thick and usually over 6 inches wide.

plug weld—A weld in a circular hole of one piece joined to a second piece in a lap joint.

porosity—Pinholes in a weld as the result of gas pockets.

postheating—The application of heat to the base metal after soldering, brazing, or welding to slow cooling.

power source—A piece of equipment supplying current and voltage for purposes of welding, cutting, or both.

preheating—The application of heat to the base metal before soldering, brazing, or welding to slow cooling.

press brake—Mechanical equipment for bending or shaping metal.

properties—Characteristics of a material such as its hardness and ductility.

quality—Welding that meets a credible standard of excellence.

rectifier—A device that converts alternating current to direct current.

regulator—An apparatus used to control the release of gas held under pressure.

root bend—Stretching the root of a coupon to test the weld.

root face—An edge that has not been removed by a bevel, establishing the boundary of the root opening.

root opening—See *joint root opening.*

root penetration—The depth of penetration by the weld into the joint.

rust—Iron oxides that form on iron and steel as the result of corrosion.

safety glasses—A critical piece of safety equipment with side shields to protect the eyes from damage relating to welding. Meets ANSI Z87.

self-shielded—Does not require an external cylinder of shielding gas.

semiautomatic—A term used to describe welding by equipment that is automatically controlled in some way.

shear strength—Resistance of a material from two equal forces in the opposite directions.

sheet metal—Metal that is 3/16 inch or less in thickness and over 6 inches wide.

shielded metal arc welding (SMAW)—An arc welding process that uses an electric arc, usually between a flux covered consumable electrode and the weld pool.

shielding gas—A gas used to protect the weld pool from contaminants in the air.

short circuiting transfer—A gas metal arc welding process in which a solid wire is extinguished by the weld pool only to reignite and begin again, up to 200 times per second.

side bend—Stretching the side of a coupon to test the weld.

silicon—A nonmetallic element, which, when added to steel, can increase its hardness and improve its mechanical properties.

slag—A nonmetallic by-product of FCAW that can affect weld quality if not removed entirely before completing another pass.

slag inclusion—Nonmetallic material trapped within the weld.

slope—A fine adjustment on some power sources that can reduce spatter.

slot weld—An elongated plug weld.

soldering—A group of processes in which the filler metal is heated to liquidus below 840°F and below the melting temperature of the base metal.

spatter—Molten metal from the electrode thrown from the weld pool.

specification—Detailed information used as a guideline for describing something.

spool gun—A device that carries wire (up to 2-pound spools) which is pulled through by drive rolls.

spot weld(ing)—A weld made on or between two lapping pieces of metal.

spray-arc transfer—A gas metal arc welding process in which the molten metal drops change to spray.

square-groove weld—A weld that uses a butt joint without edge preparation.

stainless steel—A high-alloy steel with at least 12% chromium. Stainless steel is well suited for corrosive applications.

steel—See *low-*, *medium-*, *high-*, and *very high-carbon steel.*

stress—Forces acting on a material.

stringer—A weld bead deposited without weaving.

strip—Metal under $\frac{3}{16}$ inch thick and usually not exceeding 6 inches in width.

structural shapes—The forms in which metal can be purchased.

s`urfacing weld—A weld applied to build up the surface of base metal that is subject to wear.

tack weld—A short, temporary weld used to hold a weldment in position to complete welding.

tee joint—A joint formed with two or three pieces that are usually at right angles to one another.

temper(ed)—A heat treatment by which hardened steel is toughened.

tensile strength—Resistance to being pulled apart.

tensile test—A test used to measure the strength of a material subjected to a force trying to pull the material apart.

toughness—The resistance to fracture from a constant force.

transformer—A power source that uses alternating current as the output for welding.

transverse cracking—A defect or discontinuity that occurs in the weld from weld toe to weld toe.

travel angle—The angle of the rod, electrode, gun, or torch in relation to the direction of travel along the joint.

travel speed—The movement of weld pool across the joint.

tubular—A hollow, continual wire filled with a granular flux.

tungsten—An element added to steel to increase hardenability.

ultraviolet light—Invisible light that can damage the skin and is especially harmful to the eyes.

underbead cracking—A defect that can occur in high-carbon and alloy steels in the heat-affected zone.

undercut—Groove cut into the base metal during welding, unfilled by deposited metal.

uphill—Vertical position, upward welding against gravity.

vertical position—A position in which the plane of welding is from 15° to 100° with 10° of pivot including uphill and downhill.

very high-carbon steel—A steel that contains 0.65% (0.0065) to 1.5% (0.015) carbon.

V-groove weld—A weld in which the edges of both pieces forming the joint are beveled.

volatile—Capable of exploding.

voltage—A measurement of the force causing the flow of electricity.

weaving—Any one of several patterns of movement from side to side during welding.

weld—The melting together of the joint with or without filler material.

weldability—The degree of difficulty by which a metal can be welded successfully.

weld bead—A weld from a single pass.

welder—A person with the skill and knowledge to make quality welds.

weld face—The surface of the weld, extending from the weld toe on one side to the weld toe on the other side.

welding—Any process that joins materials by heat.

welding arc—A controlled electrical discharge in which the electric current causes intense heat in a weld pool between an electrode and a base metal.

welding gloves—Gloves designed specifically for welding—not ordinary work gloves. Gauntlets protect the lower arms.

welding position—Can be flat, horizontal, vertical, or overhead.

welding procedure—Detailed instructions that must be followed to achieve quality welding.

weld leg—Measurement of a fillet weld from joint root to the weld toe. Each fillet weld has two weld legs.

weldment—An assembly of welded parts.

weld pool—Liquid metal formed at the point of welding.

weld root—The point at which the weld penetrates into the joint root; the farthest point from the weld face.

weld toe—Runs the length of the weld to establish the boundary of the weld face with the base metal.

work angle—The angle of the rod, electrode, gun, or torch in relation to the base metal.

workpiece connection—Connects the base metal to the power source; has replaced nonstandard term *ground.*

workpiece lead—A cable connecting workpiece connection to the power source.

Index

AC. *See* Alternating current
Active gases, 132
Air carbon arc, 57
Alloy steels, 49, 81
Aluminum, 28
  electrodes, 131
  expansion, 139–140
  identifying, 54
  shielding gas, 70
Alternating current, 7–9, 83–84
American Society for Testing and Materials (ASTM), 55
American Welding Society (AWS), 31–32, 89, 117, 126, 129–130, 158
Amperage, 7, 24–26, 77, 79
Angle, 153
Arc. *See* welding arc
Arc blow, 78, 83–85
  troubleshooting guide, 85
Arc flash, 34, 37–38
Arc length, 5, 6, 24, 84
Arc strike, 36, 65
Argon, 32, 71, 126–128, 132–134

Backhand welding, 66, 92–93, 97
Backing material, 116
Band saw, 57–58
Bar, 153
Base metal, 1, 2, 81, 84
Bevel, 154
Bevel-groove weld, 114, 116
Birdnesting, 26–27
Boron, 51
Brass
  identifying, 55
Brazing, 1
Buildup, 75
Butt joint, 98, 113–115

Cables
  power cables, 28–29
Cap, 39
Carbon, 51, 81
Carbide precipitation, 51
Carbon dioxide, 32, 70–71, 126–128, 132–134
Carbon steels, 5, 51, 53, 103, 130
  electrodes, 131
  specifications for electrodes, 32
Cast irons, 49, 53
  expansion, 140
Channel, 153
Chromium, 51
Clothing
  protective, 38–39
Coefficient of linear expansion, 139–140
Complete joint penetration, 98–100, 102, 114–116
Compression strength, 50
Concave contour, 117–118
Conductor tube, 28–29
Constant current, 5
Constant potential, 5
Contact tube, 3, 4, 5, 25, 77, 91
Contraction, 139–140
Convex contour, 117–118
Copper, 51
  electrodes, 131
  expansion, 140
  identifying, 55

Corner joint, 57, 113–114
Coupons, 157–161
Cracking, 81–83, 139
  troubleshooting guide, 85
Crater, 78
  cracks, 79, 82–83
Current, 7

DC. *See* Direct current.
DCEN, 9, 84, 128–129
DCEP, 9, 84, 128
Defect, 82
Deoxidizer, 126
Destructive test, 157–161
Diesel engine, 20–21
Direct current, 7–9, 83–84
Direct current electrode negative. *See* DCEN
Direct current electrode positive. *See* DCEP
Discontinuity, 82
Distortion, 51, 97–98, 140–145
Double-groove weld, 120
Downhill, 81, 100–102
Ductility, 50
Duty cycle, 18–19
  guns rated, 29

Earplugs, 22, 34–35
Edge joint, 113
Effective throat, 142–143
Electric theory, 7
  Cycle, 7–8
Electrode (wire), 1, 3, 59–60, 81–82, 125–134
  alloy steel electrodes, 128
  carbon steel electrodes, 128
  controlled, 6
  diameters, 5
  identification, 60
  sizes, 129–130
  types, 126–129
Electrode extension, 65, 67–68, 99
  troubleshooting guide, 85
Elements, 53–54
Expansion, 139–140

Fabrication, 150–151
Face bend, 157–161
Fatigue strength, 49
Feeder system, 23, 26–27, 32
  burnback control, 26
  gas purge, 26
  timer, 26
  wire reel braking system, 26
Ferrous, 49
Filler metal, 1, 2, 5, 23, 31–32, 82
  solid wire, 31
  tubular wire, 31
Fillet weld, 95, 114, 117, 142–143
Filter lens, 36–38
  filter lens chart, 37
Fire extinguishers, 13–15
  type A, 13–14
  type B, 13–14
  type C, 13–14
  type D, 13–14
Flammable materials, 14, 156
Flat bar, 57
Flat position, 60, 89–90
Flowmeter, 26, 32, 46
Flush contour, 117, 142–143
Flux, 59–60
Flux Cored Arc Welding (FCAW), 1, 5, 75, 77
  current, 9
  definition, 3, 5
  electrode extension, 67–68
  exercises, 108–110
  learning, 6
  methods, 70–71
  parameters, 134
  weather conditions, 125
Flux cored wires, 126–127
Forehand welding, 67, 92–93, 97
Frequency, 7

Gas Diffuser, 28
Gas Metal Arc Welding (GMAW), 5, 75, 77
  automatic, 4
  current, 9
  definition, 1
  electrode extension, 67
  learning, 6
  semiautomatic, 4
  weather conditions, 125
Gas purge, 26–27
Gas shielded, 70–71
Gas shielding, 3, 26, 78
Gasoline engine, 20–21
Globular transfer, 16, 69
Grain structure, 139
Grapes, 76–77
Grinder, 57–58, 115, 152–154
Groove angle, 115–116
Groove face, 115–116
Groove welds, 114–117
  double, 120
  limited thickness, 159
  single, 120
  unlimited thickness, 159
Guided bend test, 157–161
Gun, 23, 27–30
  air-cooled, 29
  insulators, 28, 67
  major parts, 5, 23–30
  on/off trigger, 27

spool gun, 28
troubleshooting guide, 85
water circulating system, 30
water-cooled, 29–30

Hardenable steels, 81
Hardness, 49, 51, 81, 139
Heat-affected zone (HAZ), 81–83, 139
Helium, 32
High-alloy steels, 51
High carbon steel, 53–55
High-frequency, 34
Horizontal position, 62, 89–90
Hot-rolled steel, 151
Hydrogen, 81

Impact strength, 49
Incomplete fusion, 78–80
lack of fusion, 85
Incomplete root penetration, 77, 85
Inductance, 24–25
Inert gases, 132
Infrared, 36
Input current, 8
Insulators, 28, 67
Intermittent welding, 51, 96
Interpass temperature, 81
Inverter, 20–21

Jig design, 158
Joint, 1
Joint design, 113–120
selection, 119–120
Joint root, 117

Keyhole, 102

Lap joint, 57, 95, 113–114
Layout, 154–156
Lead, 30
Liner, 28, 79
Longitudinal cracking, 82–83
Low alloy steels, 5, 51, 139
electrodes, 131
specifications for electrodes, 32
Low carbon steel, 53–55, 82, 139
identifying, 54–55

Magnesium
electrodes, 131
Manganese, 51
Manual welding, 21
Martensite, 139, 143
Medium carbon steel, 53–55, 81, 139
Melt-thru, 7, 69, 79–80, 85, 99, 115, 132
Metal cored wires, 126–129
Mild steel, 53, 70, 130
Mill scale, 57
Molybdenum, 51
Motor generator, 21–32

National Electrical Manufacturers Association, 18–19
Nickel, 51
electrodes, 131
Nitrogen, 2, 71, 125, 132
Nonferrous, 49
Nozzle, 3, 4, 5, 28, 29
clogged, 78

Output, 5
Overhead position, 89
Overlap, 78
Oxyacetylene torch, 57, 115
Oxygen, 2, 71, 126, 132

Parameters, 6, 24–26
Pass. *See* weld bead
Pipe, 50, 142, 151, 153
Plasma cutting torch, 57
Plate, 5, 56, 115
Plug weld, 117–119
Porosity, 78, 85
Postheating, 53, 139
Powdered metal (flux), 3
Power source, 16–26
choice of, 16–18
definition, 15
rating, 18
safety, 60
single-phase power, 16
three-phase power, 16
Preheating, 53, 81–82, 139, 143
Properties of metal, 49–50

Quality welds, 6, 7, 80

Rectifier, 8–9, 20
Regulator, 32, 46
Repair, 149–150
Respirators, 35–36
Root bends, 157–161
Root crack, 83
Root face, 115–116
Root opening, 115–116
Root penetration
incomplete, 7
Rust, 30, 57

Safety, 34–38
breathing, 130
cap, 39
clothing, 38–39
earplugs, 22, 34–35
fire extinguishers, 13
glasses, 22, 34, 81

heat, 13–15
high-frequency, 34
light, 13–15
pressurized cylinder, 32–33, 134
reminders, 42–43
respirators, 36–37
shock, 13–15
smoke, 13–15
sparks, 13–15
storage, 130
tongs, 38–39
welding gloves, 38
welding helmet, 36–38
Safety glasses, 22, 34, 81
Self-shielded, 3, 71–72
Shear strength, 49, 120
Sheet metal, 5, 56, 115, 126
Shielded Metal Arc Welding (SMAW), 5, 6, 19, 70, 73
Shielding gas, 2, 3, 4, 23, 32–33, 131–134
function, 32
Outdoor welding, 5
regulator/flowmeter, 32–33
Short circuiting transfer, 16, 68–69
electrode extension, 67
exercises, 91–103, 161–166
parameters, 132–133
Side bends, 157–161
Silicon, 49
Single-groove weld, 120
Slag, 32, 71, 79–81, 84
cooling, 4
inclusion, 80
molten, 4
removal, 5
Slope, 24–26
Slot weld, 118–119
Soldering, 1
Solid wires, 126
Spatter, 24, 26, 83, 85, 94, 132
definition, 77
Specifications, 31–32
Spray-arc transfer, 16, 70–71,
exercises, 103–107, 163–164
parameters, 133
Square-groove weld, 109, 114–116
Stainless steels, 5, 143
electrodes, 131
expansion, 140
melting point, 54
shielding gas, 70–71
specifications for electrodes, 32
Steel, 49, 82, 139
expansion, 139–140
hot-rolled, 151
melting point, 54
Steelmaking, 51–52
Stress, 65, 82, 120, 141–142
Stringer, 75–76, 82
Strip, 57–58
Structural shapes, 49–50, 151–153
flat bar, 57–58, 153
pipe, 50, 142, 151, 153
plate, 5, 56
sheet metal, 5, 56
strip, 57–58
tube, 153
Surfacing weld, 92

Tack welds, 51, 84, 94–96, 140, 143–144
Tee joint, 57, 90, 113–114
Tensile strength, 49–51, 139
Titanium
electrodes, 131
Tongs, 38–39
Tools
adjustable wrench, 40
side cutter, 39
slag hammer, 40–41
wire brush, 40–41
Tool steels, 51
Tools, 144, 151–156
Toughness, 50, 139
Transformer, 19–20
Transverse cracking, 82–83
Travel angle, 65
troubleshooting guide, 85
Travel speed, 51, 68, 77–78, 80–81, 98–99, 101, 143
troubleshooting guide, 85
Troubleshooting guide, 85
Tube, 153
rectangle, 153
square, 153
Tubular wire (electrode), 1, 3, 4
Tungsten, 51

Ultraviolet light, 13–14, 128
Underbead cracking, 54, 81
Undercut, 67, 76–80, 85
definition, 77
Uphill, 75, 89

Ventilation, 41
Vertical position, 75–77, 89
V-groove weld, 107, 114–116
Voltage, 5, 6, 78, 91
definition, 7
adjustment, 24–25
troubleshooting guide, 85

Water circulating system, 30
Weave, 75–77
Weld, 1, 2, 77, 79, 81–82,
rust, 152
sizing, 119–120
Weld bead, 5, 75–78, 82–83, 94
Weldability, 51

Welder, 5, 7, 149
  decisions, 65–66
Weld face, 77, 117–118
Welding, 77, 149–150
  application, 3
  idefinition, 1
Welding arc, 4, 65–66, 78, 80, 85
  arc blow, 83–85
  definition, 1, 3
  instability, 9, 132
Welding gloves, 38
Welding helmet, 36–38
Welding procedures, 81
Welding technique, 71
Weld legs, 117
Weldment, 65, 94, 98, 151
  improper preparation, 77, 116
Weld metal, 3, 4
Weld pool, 1, 3, 4, 5, 69, 77–78, 102, 125
Weld root, 142–143
Weld toes, 77–78, 82, 117
Whiskers, 85
Wire. *See* Electrode
Wire feed speed, 24, 26, 77–80, 91
  troubleshooting guide, 85
Work angle, 66–67, 79, 105, 165
  troubleshooting guide, 85
Workpiece connection, 9, 30–31, 84
Workpiece lead, 30